PRAISE FOR
Sleepless in America

"A truly impressive book about a very important issue. Kurcinka skillfully acquaints the reader with research indicating the link between sleep and behavior difficulties and offers realistic, practical solutions. Her appreciation and empathy for the challenges of parenting are evident on every page of this remarkable book."

—Robert Brooks, Ph.D., Harvard University, coauthor of *Raising Resilient Children*

"Just when you thought you'd read every sleep book out there, along comes this virtual sleep support group. Highly recommended for public libraries."

—*Library Journal*

"Mary Sheedy Kurcinka really understands children. *Sleepless in America* helps parents care for their children in ways essential to child raising, but ways parents might not have thought of themselves. The practical strategies in the book resonate completely with today's family life, and the science-based information has immediate applicability. Mary's writing style is easy to read, as always. *Sleepless in America* is a powerful resource for any family, and especially families whose children are a challenge for parents and schools."

—Michael Gurian and Kathy Stevens, coauthors of *The Minds of Boys: Saving Our Sons from Falling Behind in School and Life*

"Mary Sheedy Kurcinka insightfully homes in on the details of family routines that can contribute to (or undermine) getting a good night's sleep, offering memorable, engaging family anecdotes and 'antidotes.' Mary's book is an excellent and unique addition to the available parenting literature focusing on sleep and sleeplessness!"

—Laurel Wills, M.D., FAAP

"A welcome and innovative guide for both parents and health care providers. I am excited to incorporate its wisdom into my own pediatric practice."

—Marjorie Hogan, M.D., University of Minnesota

"Even the most seasoned parent can be pushed beyond the breaking point by a child who won't sleep. No matter how patient we are with our children during the day, nights with tired, cranky children can bring us to our knees. Up against our own exhaustion, we find ourselves powerless to help our children find the deep, sustaining rest they need. In this wise, illuminating book, Mary Kurcinka unlocks the secrets to a good night's sleep for parents and children alike."

—Laura Davis, author of *Becoming the Parent You Want to Be* and *I Thought We'd Never Speak Again*

"In her new book on sleep problems, Mary Sheedy Kurcinka again demonstrates her extensive grasp of the nature and needs of today's children. A great many parents will find this book helpful in its comprehensiveness, accuracy, clarity, and good sense."

—William B. Carey, M.D., The Children's Hospital of Philadelphia

"This book provides a compelling collection of stories and advice on children's sleeping. The advice follows simple principles, but takes account of the many differences between children and their families. The book is directed at parents, but practitioners will also find many useful ideas."

—John E. Bates, Ph.D., Indiana University

"Kurcinka addresses topics such as morning and evening bedtime routines, naps, holidays, and nightmares with practical, easy-to-read advice. Her real-life examples come from thousands of parents in the classes and workshops she leads around the country."

—*Seattle Times*

About the Author

MARY SHEEDY KURCINKA, M.A., has more than twenty years' experience as a teacher, an award-winning parent educator, and an international trainer. The founder of the Spirited Child and Power Struggles workshops, Kurcinka is also the bestselling author of *Raising Your Spirited Child* and *Kids, Parents, and Power Struggles*. She lives in Minnesota.

Also by Mary Sheedy Kurcinka

Raising Your Spirited Child

Kids, Parents, and Power Struggles

Raising Your Spirited Child Workbook

SLEEPLESS
— IN —
AMERICA

Is Your Child Misbehaving
. . . or Missing Sleep?

Mary Sheedy Kurcinka

HARPER

NEW YORK • LONDON • TORONTO • SYDNEY

HARPER

A hardcover edition of this book was published in 2006 by HarperCollins Publishers.

HarperCollins books may be purchased for educational, business, or sales promotional use. For information please write: Special Markets Department, HarperCollins Publishers, 10 East 53rd Street, New York, NY 10022.

First Harper paperback published 2007.

Designed by Joseph Rutt

The Library of Congress has catalogued the hardcover edition as follows:
Kurcinka, Mary Sheedy
Sleepless in America : is your child misbehaving or missing sleep? /
Mary Sheedy Kurcinka.
p. cm.
1. Sleep disorders in children—Popular works. 2. Children—Sleep—Popular works.
3. Child rearing. I. Title.

RJ506.S55K87 2005
618.92'8498—dc22 2005046179

ISBN: 978-0-06-073602-6 (pbk.)
ISBN-10: 0-06-073602-X (pbk.)

07 08 09 10 11 ❖/RRD 10 9 8 7 6 5 4 3 2

Dedicated to my father, Richard Dennis Sheedy,
who loved to arise early in the morning to tend his dairy farm,
always took his ten-minute nap,
and modeled for his five daughters a life of balance,
laughter, and commitment to family

ACKNOWLEDGMENTS

Writing *Sleepless in America* has been a journey that has taken many years. I could not have completed it without the support, guidance, and insights of many others. I am very grateful for their contributions and wish to say thank you. In random order as always . . .

All of the parents, and children who have shared their stories with me including their successes and tribulations, I have learned so much from you.

Joseph M. Kurcinka, my husband, favorite visionary, sounding board, life partner, and creator of this title.

Joshua T. Kurcinka, my son and marketing expert.

Kristina L. Kurcinka, my daughter, who patiently read chapters, providing witty, sometimes pointed, and amazingly perceptive feedback.

Betsy McAllister, my daughter-in-law, the computer whiz who has provided essential technical assistance.

Patti Manolakis, parent, writer, and guide, who has stuck with me through every single word, from the proposal to the final manuscript.

Dr. Laurel Wills, MD, who encouraged me and provided key technical guidance throughout the developmental process.

Dr. William Carey, MD, passionate pediatrician and temperament researcher.

Sara Harkness, Ph.D., University of Connecticut, cross-cultural temperament researcher and very helpful guide.

Jim Cameron, Ph.D., Preventive Ounce, Berkeley, California, wise, wonderful, temperament guru.

Laura Davis, fellow writer and advocate for children.

Ginger Sall, LaLeche League Leader and mother who was willing to share her wisdom.

Charles Super, University of Connecticut, positive, supportive guide.

All of the other researchers who were willing to share their findings and insights with me, including: Jay Belsky, John Bates, Joanne Cantor, Mary Carskadon, Sue Carter, Ronald Chervin, Ron Dahl, Nathan Fox, John Gottman, Megan Gunnar, Marie Hayes, Heather Henderson, Pamela High, Jerome Kagan, Barbara Keogh, Roy Martin, James McKenna, Claire Novosad, Judith Owens, Steven Porges, Mary Rothbart, Robert Sapolsky, Judith M. Stern, Robert Thayer, and Erik Turkheimer.

Beatrice Sheedy, my mother, provider of emotional support even after all these years!

Barbara Majerus, Kathy Kurz, Helen Kennedy and Suzanne Nelson, my sisters who each in her own way encourages me, and keeps me grounded.

Mary E. Kurcinka, my mother-in-law, who graciously handles all the "wrong numbers" that come to her.

Kim Cardwell, friend and colleague who has been there for me, with every single book.

Lynn Jessen, friend and colleague, owner of Paidea Child Development Center, who makes every decision by asking what's best for children.

Pam Kennedy, a very wise mother, grandmother, and nurse who understands the needs of children, especially infants.

Marietta Rice, friend, colleague, and unending source of creative ideas.

Jenna Ruble, friend and colleague, amazing teacher of young children, and mother of three.

Gail Winston, my insightful editor whose guidance has been so helpful.

Katherine Hill, assistant to Gail, the conscientious caretaker of details.

Heide Lange, my agent, champion of my work, and dream maker.

Gary Smolik, fellow writer and amazing wordsmith.

Susan Perry, teacher and writer, who helped me to finally get the proposal for this book right!

Dawn Kalb, ever dependable assistant, keeper of my office and daily details.

Wendy Kennedy, mother of three, adviser, and guide.

Michael and Christine Blonski, advisers and storytellers.

The teachers of Paidea, outstanding staff, and experts at napping 125 children.

Harry and Francy McAllister, collectors of great stories, skilled teachers, and parents.

Dr. Margie Hogan and Dr. David Griffin, friends and brilliant physicians, who introduced me to Dr. Laurel Wills.

Scott and Susan Wills, warm, wonderful supporters.

Tom, Lindsey, and Kellen Fish, neighbors, friends, and key players on my support team.

Bob, Jane, Samantha, and Danielle Herr, my delightful neighbors, who greet me with smiles every morning and are willing to "test" my theories.

Brittany Basye, mother of three and coach of many.

Raelene Ostberg, student teacher, now parent educator, who asked the best questions.

Mary Jo Cole, mother, musician, and storyteller.

Betty Cooke, Ph.D., University of Minnesota, energetic leader in Minnesota's parent education programs.

Ed and Sylvia Kwan, Jeff and Dawn Ellerd, Betsy Preus, Nicole Gilliand, Robert Green, Ruth Dietz, and Terri Skadron, wonderful supporters willing to spend time giving me feedback.

Linda Bruhn and Jessica Page, organizers extraordinaire, who keep my home and office sparkling and give me time to write.

Joanne Burke, loyal friend and supporter.

And not to be forgotten, Bridger, my eighty-pound black Labrador, who lies at my feet and offers me comfort as I write.

Thank you! Thank you! Thank you!

CONTENTS

Introduction xv

SEEING THE LINK BETWEEN MISBEHAVIOR AND MISSING SLEEP

ONE Temper Tantrums, Morning Wars, Homework Hassles 3
What does sleep have to do with misbehavior?

TWO Is Your Child Misbehaving or Missing Sleep? 18

THREE Short-Tempered, Feeling Overwhelmed: Are You Missing Sleep? 28
How your fatigue affects your child's behavior

LEARNING ABOUT SLEEP FROM A NEW POINT OF VIEW

FOUR How Do You Get Your Child to Sleep? 41
Sorting through all of the advice

FIVE Why Kids Do or Do Not Sleep 57
It's a matter of physiology, not a power play

EXPOSING THE CULPRITS THAT CAN KEEP YOUR CHILD AWAKE

SIX Tension Triggers 69
Is your child too tense to sleep?

SEVEN Easing the Tension 85
Helping sleep come quickly

EIGHT Time 102
Recognizing what upsets your child's body clock

NINE A Good Night's Sleep Begins in the Morning 121
Setting the body clock

TEN Temperament 145
Is your child's genetic "wiring" making it more challenging to sleep?

ENLISTING EFFECTIVE STRATEGIES FOR SOUND SLEEP AND GOOD BEHAVIOR

ELEVEN Ending the Bedtime Battles 167

TWELVE Customizing the Bedtime Routine to Fit Your Child 188

THIRTEEN Night Waking, Night Terrors, and Nightmares 205
Quieting the screams in the night

FOURTEEN Naps and Siesta Time 229
Eliminating the late-afternoon meltdowns

PREVENTING POTENTIAL PROBLEMS

FIFTEEN Infants 251
Getting Off to a Good Start

SIXTEEN Taking the Fight Out of the Morning Start 267
Adjusting wake and bedtimes

SEVENTEEN Travel and Holidays 283
Planning for success

EIGHTEEN Changing Beds 296
Moving out of your bed, out of the crib, or from one bed to another

THE JOYS OF SOUND SLEEP AND GOOD BEHAVIOR

Conclusion 315

Index 317

INTRODUCTION

It was a four-year-old child who first piqued my interest in sleep. Her parents had invited me to their home because of her tantrums. The outbursts were loud, vehement, and unremitting. During that late-afternoon meeting, as I sat at the kitchen table talking with the parents, she climbed up on the couch and fell soundly asleep. At the time, I was focused on the "issues," morning meltdowns, bedtime battles, late-afternoon sibling squabbles. But for some indiscernible reason, her soft, steady breaths caught my attention. And I asked her parents how much sleep she was getting. I learned that she slept eight to nine hours a night, but when I suggested that she might need more sleep, they were skeptical about the idea. "I try to put her to bed," her mom told me. "It's a dragged-out battle. She doesn't sleep. I'm beginning to think she's one of those kids who just doesn't need much sleep."

Yet the tried-and-true strategies to prevent the tantrums which I was suggesting were not working with this family. I was puzzled and intrigued. Intuitively, I sensed that this child was exhausted, and yet I didn't understand why, if she was so tired, she was vehemently refusing to sleep. It was a mystery I couldn't resist.

I began reviewing the research, and discovered a burgeoning body of information confirming the scientific link between misbehavior and lack of sleep. I also found out that overtired children are often too tense to sleep. Excited by what I'd learned, I began contacting the researchers themselves, and found them not only open but eager to talk with me about their work. They shared their studies and results with me, and connected me to their network of colleagues. One interview led to another, until the pile of notes on my desk threatened to bury me.

And then I began to listen carefully to the questions asked by parents and professionals who attended my classes and workshops. I quickly realized that I was not the only one who had missed the link between challenging behavior and lack of sleep. In fact, it was a silent epidemic, mysteriously taking its toll not only on the children, but on their entire families as well. Sleep, or lack of it, I realized, had the potential of revolutionizing how we as parents and as a society dealt with "challenging behaviors."

By the way, I'm Mary, licensed teacher of parents and children. Yes, in Minnesota, where I live, one is required by state law to have a degree in parent education and a license issued by the Department of Education, in order to work with families. I've been providing educational seminars and one-on-one consultations with families for nearly three decades.

Eager to share what I was learning, I began adding bits and pieces of the information to my ongoing "Raising Your Spirited Child" and "Kids, Parents and Power Struggles" classes, but I soon realized I needed to develop an entirely new curriculum, one solely focused on sleep, which not only addressed the importance of it but helped families to tackle the dilemma they were facing—how to get enough sleep in a twenty-four-hour culture that makes overwhelming demands on their time. I wanted to give them a guide that offered both support and effective strategies, techniques they could customize to "fit" their child and their lifestyle.

To my delight, it was a topic of interest to many others, and the classes took off. Meeting together, we shared our concerns, frustrations, and challenges. I brought in not only the most renowned sleep research, but also the latest studies on temperament, stress, and child development. We analyzed and debated the material, deciphering the most helpful and practical methods to find time to sleep and to get resistant children into bed.

We laughed together, and shared the "secrets" of sound sleep for us. Often we were incredulous at our differences. What sent one person soundly off to sleep, left another wide awake in the night. Some loved to cuddle together, often in one big bed, while others found separate beds and bedrooms to be their preference.

By sharing our stories, we allowed each other to peek into our homes and even our bedrooms. We discovered that we were not the only ones who had lied to the pediatrician rather than admit our kids' true sleeping habits.

Amazingly, as we found ways to help our families get more sleep, the effect also became apparent during the day. The morning meltdowns disappeared. Late-afternoon sibling battles diminished drastically, and bedtime became an event to look forward to rather than one to dread.

As word spread about my interest in the connection between lack of sleep and poor behavior, I started receiving e-mails from all over the world, from tired parents who, like those in my classes, suspected that their children were exhausted yet found the answer to be more complex than simply putting them to bed. The result is this book.

Sleepless in America is based on the most up-to-date scientific research but, like my classes, it's not just the theories. Also included are stories that reveal challenges faced by REAL families. I've collected them from the thousands of parents in my classes and workshops, and sorted through them to find patterns. Included are the common questions and concerns, most effective techniques, and enough "horror stories" to let you know you are not alone in your struggle to help your family get more sleep.

As we go along, you'll discover that a good night's sleep does not just begin at bedtime. A good night's sleep begins in the morning. By becoming aware of decisions that affect sleep for ourselves and our children, we have the power to make a difference. While we cannot MAKE our children sleep, we can create an environment that values sleep and is conducive to it.

Sleep is dynamic, ever-changing, and keenly sensitive to the stresses and strains of daily life. I promise you, we can make it much better. I cannot promise you perfection. There will still be an occasional "bad night." But it's much easier to endure when you know how to make it better—next time. And, gradually, as you become more astute at recognizing those seemingly innocent things that are undermining your family's sleep, the good nights really do begin to roll into one another. And, as they do, the daily power struggles begin to disappear, slipping away with the soft breaths in the night. When we are well rested, we are more open to listening to one another, more patient, and we even find each other's company much more enjoyable.

So, grab your pillow, find a comfortable chair, and begin to discover how to stop those daily conflicts by choosing sleep.

And oh, by the way, that little girl—she's regularly sleeping twelve hours a day now, and the tantrums are so infrequent that it's difficult to remember the tough times.

SEEING THE LINK BETWEEN MISBEHAVIOR AND MISSING SLEEP

ONE

TEMPER TANTRUMS, MORNING WARS, HOMEWORK HASSLES

What does sleep have to do with misbehavior?

"The difference between a child who is well rested and one who is not is a smile on his face—and on yours."

—Joe, father of two

The trouble with a child who is missing sleep is that her behavior is confusing. It's hard to believe that the real culprit behind her temper tantrum is lack of sleep when bedtime is one of your biggest battles, or she loses it simply because you dropped her water bottle. And when she can't even dress herself, even though she did it yesterday, it feels more like a plot against you than an issue of fatigue. How can a child who is supposedly so tired somehow garner the energy to veer off her path just far enough to bop her brother in the head, and jump on her bed laughing hysterically when you try to get her down for the night?

But if your child is misbehaving, it's very likely that he or she is crying for sleep. Sleep-deprived children can include babies who are sleeping less than 14–16 hours in a 24-hour period; toddlers sleeping less than 13 hours, preschoolers less than 12 hours, school-age children less than 10 hours, or adolescents sleeping less than 9.25 hours a night. And until your child gets more sleep, no punishment, no discipline strategy will stop the challenging behaviors. Sound sleep is a key to good behavior. The problem is that children rarely tell you that they are tired. Instead, they get wired, which escalates into a frenzy of energy. It's as though their body is out of control—and it is.

Suspecting that your child might be tired, you may have even tried to put him to bed at a reasonable hour, but it's as though he fights sleep. If he's an infant, just as you think he is about to drop off, he jerks awake, thrashing and shrieking. And if he is older, no matter what you do, he still complains that he can't fall asleep, wakes frequently in the night, and all too often awakens early. Since your efforts are unrewarded, it's easy to assume that he does not need much sleep. The misbehavior and whining continue, and the connection to lack of sleep remains a mystery. That's what happened in Samantha's family.

On Saturday, eight-year-old Samantha was a delight. She accepted the news that her favorite cereal was gone with a mere sigh of disappointment. Over breakfast, she chatted cheerfully with her parents and even allowed her brother to join the conversation. When the baby reached for her toast, she offered him a bite instead of slapping his hand. He squealed with pleasure. Without complaining, she cleaned her room, and didn't lag behind on a trip to the shopping center. Her parents grinned, proud of their skill and glorying in their daughter's energy and enthusiasm for life. But Sunday was a different story.

On Sunday, Samantha wouldn't get out of bed, despite the planned outing with her grandparents. She shrieked in protest when her mother announced it was too cold to wear shorts, and shoved her brother away when he came to investigate the problem. The baby, hearing the high-pitched screams, sat saucer-eyed in fear. Unfortunately, it was also a bad-hair day, an occurrence that overwhelmed Samantha and dropped her to her knees, tears spurting from her eyes. No matter what her parents did to remedy the situation, they couldn't get it right. She reeled in their arms, and then bolted from the room. Same child, same parents, same week—why such a difference in mood and behavior?

On Friday night, Samantha had enjoyed ten hours of sound slumber. She had been so pleasant on Saturday that her parents rewarded her by letting her stay up late to watch a movie. But on Sunday morning, plans precluded her sleeping in, leaving Samantha short on sleep. The tantrum got her parents' attention but not the association with lack of sleep.

SOMETIMES IT'S EASIER TO RECOGNIZE SLEEP DEPRIVATION IN YOURSELF

I first met Samantha's mother, Sara, when she attended one of the weekly classes I teach for parents in St. Paul, Minnesota. Every week, for eight weeks, sixteen parents and their children arrive at the center.

From the beginning, some stroll into the room ready to visit with friends, and learn new, effective strategies for working together. Others initially slip quietly through the door, weary. They wonder if there really is information that can help them, or if they are the only ones facing the issues that trouble them. Almost always I am rewarded weeks later, as they, too, arrive smiling, with a proud stride to their step. I never grow tired of welcoming them, and am always eager to understand the issues they face.

When I'm not teaching in St. Paul, I lead large workshops all over North America, as well as offer private consultations for families. I also write. My previous books have included *Raising Your Spirited Child; Raising Your Spirited Child Workbook;* and *Kids, Parents and Power Struggles.* Whether I am writing, working one on one with a family, leading a small discussion group, or speaking to a thousand parents and professionals, I am always deeply grateful that this is my work.

Initially, when I brought up the topic of sleep in class, Sara was skeptical. What could sleep—or lack of it—possibly have to do with the fact that Samantha was constantly overreacting to the simplest requests? Or that her four-year-old son became "mother-deaf," unwilling to listen every afternoon at five. Especially since the mere word "bedtime" could send both of them to the moon. So, I asked her to take note of her own feelings when she was short on sleep. The next week, she had a story to tell.

"The baby had an ear infection," she explained. "I didn't recognize it for several days, because he wasn't running a temperature. He wasn't sleeping. I was up at least four times a night, and even when he did sleep, I was lying there, waiting for him to wake up again. So, when the alarm went off, I snuggled deeper under the covers, breathing slowly, my eyes closed. Just a minute, I thought. Just give me a minute. But there was no time for rest. I threw back the covers and dragged myself to the closet. I stood there, stumped. I couldn't make a decision. A dull headache thudded in the back of my head. Frustrated, I turned to leave and stubbed my toe. Searing pain shot through my entire leg. I couldn't believe how much it hurt. Now limping, I headed in slow motion to the kitchen, hoping that eating something would ease the weight in my limbs and the dryness in my mouth. Before I got there, my husband asked me to pick up the newspaper at the front door. I snapped at him. His simple request was overwhelming to me. Entering the kitchen, I opened up the bread drawer—no, I thought, toast would be too dry. I stood there, unable to figure out what I wanted. 'Aren't you eating anything?' my husband asked. I couldn't answer him. Instead, I burst into tears."

She paused, her voice dropping. "Until that day I never realized my kids must be feeling the same way."

Sara is not alone. I have to admit that in years past, as I worked with families, I, too, misinterpreted the signs of sleep deprivation. When parents asked me what they should do when an afternoon's outing to the discount store erupted into a temper tantrum merely because they replied "maybe" to a request for a cookie, or a child constantly demanded attention, I simply responded to the behavior. I didn't know better. As a result, I was frequently puzzled and embarrassed when the strategies I suggested—which were tried-and-true techniques—didn't work. I began searching for an answer to explain why, and I found it—so simple and yet so elusive—sound sleep. Today, my first question when consulting with families is: "How much sleep is your family getting?"

WHY IS MISSING SLEEP THE CULPRIT?

New research has demonstrated the key role adequate sleep plays in the ability to control one's emotions, behavior, and attention span. Without sufficient sleep, your child's performance, mood, focus, and ability to work with others deteriorate rapidly. Power struggles begin with a lousy night's sleep. Even the most compliant child starts to lose it over the "little things."

Researchers are also recognizing that during the first two years of life, the average child spends fourteen of those twenty-four months asleep. During this time, the brain has reached 90 percent of adult size, and the child has attained amazing skills. By the age of five, children have spent half of their lives asleep. There is an evolutionary reason for why we sleep this much in the early years. The reason, researchers believe, is that sleep plays a fundamental role in healthy brain development, leaving us with a pressing question: Can our children afford not to get enough sleep?

HOW DID WE GET HERE?

In the United States, there's been a huge cultural shift during the past two decades, one that has affected the amount of sleep children are getting. Today, there is growing pressure to engage ever-younger children in a dizzying variety of activities and experiences. Even infants are impacted as they are towed along to older siblings' activities and exposed to high stimulation levels that leave their bodies too tense to sleep. Long commutes and work hours for parents also mean that if there is going to be any "family time," it's likely to occur at night. Unfortunately, it's leading to behavior problems for our children.

Curiously, other cultures are aware of the problem, as Sara Harkness, a professor at the University of Connecticut's School of Family Studies, has discovered in her cross-cultural studies of families. The Dutch believe their children need rest in order to maximize growth and development. They make sleep a priority, and therefore protect it.

Sara Harkness learned all this firsthand. While she was conducting her study, Sara's own children were enrolled in a Dutch elementary school. One evening, a school program ran an hour later than the children's usual bedtime. Despite the fact that there was no announcement or written notice, every parent there—except Sara—knew that the next day, school would start an hour later, so that the children could get their sleep.

Things are very different in the United States. Recently, Kim, the mother of three, described taking her five-year-old son, Michael, to a children's concert. It was supposed to be a fun family outing, but instead it turned into a public battle. Once inside the concert hall, Michael refused to walk down the steps to his seat. Kim found herself dragging him down the aisle. As soon as the concert began, Michael proceeded to whine and complain that the lights hurt his eyes and that the music was too loud. The glowering stares of the people around them drove Kim to take Michael back to the lobby. There, he caught sight of the souvenirs. When she refused to buy him one, he threw a knock-down temper tantrum, complete with shrieking and kicking. Kim had had enough. She took him home. Once in their house, she sent him to his room. He cried for a bit, and then fell asleep on his bed. While he slept, Kim fumed. The tickets had been expensive, yet he'd been so ungrateful. He'd turned what was supposed to be a "fun" outing into a fight.

It was then that she picked up the telephone and called me. She'd heard me speak at a local seminar. Knowing that I had written books for parents and that I taught parent-education classes, she hoped I might be able to help.

Her anger quickly spilled into the conversation. She described the glares of the other concertgoers and her own embarrassment, and then asked, "How should I punish Michael for acting that way?"

I let her talk until the intensity faded from her voice and then simply asked, "What do you think he was feeling?" She groaned. "I think he was worn out," she said. "Worn out from what?" I asked. She explained that he'd recently started a new school, which was very draining for him. He was also taking karate and gymnastics lessons three nights a week. But the real confirmation of his exhaustion had been his own words. Just that morning, he'd entreated, "Mom, I'm too tired to go. Please, can't we stay home?" Her voice soft and despondent, Kim said, "I heard him, but I didn't want him to miss the opportunity." Kim and Michael are not alone. Children of all ages are simply not getting enough sleep as we frantically attempt to balance the demands and realities of life today with the needs of our children who are literally crying for sleep.

As a parent, I have lived this reality. As a parent-educator, I see it every day. I understand the struggle of trying to "fit it all in," the constant mental wrestling matches of priorities and responsibilities. How does one honor a bedtime when it means missing a school play? Is it more crucial to insist your child get sleep or participate in one more activity? Is a nap really that helpful? When a workday and commute grab ten or eleven hours a day, how does one find time for family without skipping sleep? It can feel like a lonely struggle. The reality is that you are not the only one. A recent poll conducted by the National Sleep Foundation found that 69 percent of parents reported that their children experience one or more of the sleep-related "misbehaviors."

DISCOVERING THAT YOUR CHILD IS MISSING SLEEP

You may have recognized that since infancy your child was sleeping less than other babies. Perhaps she had colic and screamed for hours, or caught twenty-minute "cat naps" on the fly. Maybe it wasn't until she dropped her nap that late afternoons became the "poison hour." Or perhaps you're not even sure that your child is missing sleep. Maybe he's been irritable and short-tempered for so long you're not certain if this is fatigue or his personality. And since the "good days" can be interspersed with the "bad ones," you may have thought he was deliberately acting this way just to "get you." What you do know is that the conflicts and temper tantrums are too frequent, the whining and constant demands exhausting, and the inability to listen and pay attention infuriating. You've tried every discipline strategy in the book, yet his poor behavior is still getting him into trouble.

THE SIGNS OF SLEEP DEPRIVATION

Sleep deprivation is confusing. It's baffling when you are out with friends, and your child, who has missed her nap, seems unfazed. Thriving on the excitement and interaction, she's crazy with energy—until you arrive home and she falls apart.

That's because the quest to stay alert is so strong for children that instead of getting drowsy, many get "wired." Their behavior appears wild rather than tired as long as stimulation levels are high enough to keep them awake. But at home, when stimulation levels drop, they torment siblings, argue with you, and chase pets, all in a mad drive to create enough commotion to stay awake.

Sleep deprivation is also sneaky. It's cumulative, and creeps up on you. Miss a nap on Saturday, watch a late movie on Sunday, attend a weeknight hockey practice that extends an hour past your child's normal bedtime, and the result is a child hours short of sleep. Yet, two days later, the whining, the shove on the playground, the inability to stay focused, the poor test score, or the fit over the dirty T-shirt don't necessarily scream out "sleep deprived" to you. All too often, however, it is the culprit lurking in the background. But it seems so insignificant. Can missing an hour or two of sleep really make a difference? Surprisingly, the answer is a resounding yes. Every child has "bad days," but when a child is fatigued, her most challenging behaviors are more rampant. She loses the ability to "regulate" her actions and responses, and, as a result, the intensity and frequency of misbehavior increase exponentially. Not all children will demonstrate all of the following behaviors, but each will exhibit enough to make you realize something is amiss. When your child is sleep-deprived, you will see more difficulty managing her emotions, her body, her focus, and interaction with others.

DIFFICULTY MANAGING EMOTIONS

Sometimes it's obvious that your child is exhausted. Little things that would never bother him on a "good day" send him over the edge. A request to put on pajamas catapults him into a full-fledged tantrum. Dad unexpectedly arriving to pick him up, instead of Mom, as expected, is cause for tears. When he "loses" it, you know it will be a bloody battle trying to bring him back. Every emotion is exaggerated. When he's not screaming, he's laughing hysterically at his brother's joke. The one that wasn't all that funny.

But sometimes it's not quite so apparent. Sleep deprivation also lowers one's ability to manage pain. The child who experiences headaches in the late afternoon may actually be short on sleep. A bump on the head can send him into orbit.

The whiny, clingy, anxious child may also be fatigued. She can't tell you what she needs, and instead slips into the words and tone of a much younger child.

All children have more difficulty managing their emotions when they are fatigued, but some children, by their very nature, demonstrate stronger emotional reactions. These children—the spirited ones—are especially vulnerable to missing sleep. They're already working harder to keep their emotions in check, slow their bodies, and calm themselves. Without sleep, the task becomes excruciatingly difficult.

DIFFICULTY CONTROLLING THE BODY AND IMPULSES

It's not that children never hit. They do, but when they're tired, they hit more frequently. As exhaustion increases, it's as though they become crazed. There's fire in their eyes, and their hands are in the snack cupboard, grabbing carbohydrates to keep going. They hit and throw things, when on a "good day" they would not. Naptime is resisted, and bedtime is a battle.

As Stephen Sheldon, a pediatrician at Northwestern University and director of the Sleep Medicine Center at Children's Hospital in Chicago, noted in a *U.S. News and World Report* article, "Tired children have difficulty regulating their behavior. One theory is that the brain and muscles in children are not synchronized and don't tire simultaneously. Instead, the brain gets tired first and loses control of the muscles, meaning drowsy kids may start running around chaotically."

It's hard to imagine that the wild child is actually tired. The clue is the "frenzy" of activity. A high-energy child is on the move throughout the day, but her movements are smooth and well coordinated. Fatigue alters this fluidity. Actions become jerky, tight, and frenzied. Unexpectedly, she trips or falls. Research has shown that preschoolers who sleep less than ten hours a day are 86 percent more likely to incur injuries requiring emergency-room treatment.

And then there is the child whose body does slow down, to the point that he can barely function. Suddenly, a child who always dressed himself simply can't. A child who carried his own book bag finds it too heavy. The child who is capable of walking needs to be carried. If sleep deprivation continues, his immune system is weakened and he succumbs to whatever "bug" is going around.

INABILITY TO STAY FOCUSED AND TO PERFORM WELL

It's the exhausted child who insists on having the radio or television turned up louder and louder. "I can't hear it," he shouts above the roar. That's because a tired child will constantly seek stimulation and change focus in order to stay alert. He gets into trouble for incessantly demanding your attention, not listening or not staying on task.

In a study of 866 children age two to fourteen, parents who said their children were sleepy also reported that the children were easily distracted and forgetful, talked excessively, fidgeted, took inappropriate actions, and had difficulty completing tasks.

The completed assignment left at home, or the missing jacket, may reflect your child's exhaustion rather than his irresponsibility. Even the ability to make decisions disappears with fatigue. Figuring out what you are feeling or what you need takes energy and focus. Without it, it's hard to decide which shirt to wear and whether or not you're in the mood for spaghetti or pizza. Exhaustion leads to irritating waffling and, ultimately, a puddle of tears.

Performance also deteriorates when attention is unfocused. Avi Sadeh, a researcher from Tel Aviv University, found that sleep-deprived children (restricted by a modest forty-one-minute shortfall) react more slowly and have diminished mathematical, verbal, and memory skills.

Children are not aware of their shrinking capabilities when they're sleep-deprived, and it's often missed by adults as well. Yet the impact on performance of one hour of sleep deprivation accumulated over eight days can be as significant as being totally deprived of sleep for twenty-four hours.

DIFFICULTY GETTING ALONG WITH OTHERS

It was six thirty p.m., and three-year-old Emma was exhausted. When her five-year-old brother Bjorn picked up the packet of books Grandma had given them, she snatched them back, screaming, "Mine! Mine! Mine!" Her mother touched her lightly and said, "Emma, you both want the books, so what else could you do?" Normally, Emma would have stopped to think about this request, but on this night, she simply declared, "NOTHING!" "Could you each have two?" her mother suggested. "NO!" Emma shouted, pushing her mother's hand away.

"Can you tell your brother when you'll be finished with the books?" she continued. "NO!" Emma shrieked.

"Could you find something else for your brother while he's waiting?" Mom asked in one last, valiant attempt to mediate the situation. But once again, Emma turned away, clutching the books and shouting "NO!"

This was not typical behavior for Emma. Usually, she listened well and solved problems easily, but on this night, fatigue had robbed her of any critical-thinking skills. She could not solve a problem, could not consider other options, or even think. She could only grab her books and shout "NO!"

Working with others requires the ability to manage emotions, remember social rules, and, at the same time, decipher information about the situation. It's a very complex process, which can become nearly impossible when a child is missing sleep. Exhaustion leads to rigidity and locking in, because to "shift attention" and think of a different solution takes energy that's not available when a child is fatigued. The result is frequent conflicts with siblings, peers, and you.

SLEEP DEPRIVATION OR ADHD

Looking at these symptoms, you might think these are common characteristics of children who have been diagnosed with attention deficit disorder. In fact, research demonstrates that perhaps as many as 20 percent of children who have been diagnosed with ADHD actually have a sleep disorder. The reality is that an estimated 69 percent of American children, from infants to teens, are short on sleep. Might yours be one of them?

YOU CAN USE THIS BOOK AS A GUIDE

Fortunately, the parents who have participated in my classes and workshops have been willing to share their experiences with you. I've now collected their stories, read the research, and interviewed the researchers. The result is this book. Like the parents in my classes, let it serve you as a friend, a guide who understands the challenges you face, the emotions you experience, and offers you practical strategies that are effective and long-lasting. Take what fits for you and leave the rest. You know your family better than anyone else. Working together, we can move toward our goals of less conflict and more sleep.

As we go along, I'd like you to know the things I believe will help you the most in becoming a well-rested family.

THE ESSENTIALS

You're not the only one

There are literally millions of parents who struggle with children in need of sleep and a schedule that doesn't seem to allow it. You are not the only one with lights on at midnight, or who is playing musical beds in a heroic attempt to get some sleep. Your child is not "spoiled" or "inflexible." You are not a "pushover." Truly, there are others who understand your frustration and exhaustion.

You can make a difference

You can't force your child to sleep, but you can create an environment that reduces tension and makes it much easier for your child to slip into sleep and to sleep more soundly. Often, when your child is misbehaving due to lack of sleep, it seems like a battle of wills that leaves you wondering what you are doing wrong and why your child is acting this way. But sleep is actually a matter of biology. Your child's brain must choose between being asleep and alert. If his body is tense, he's on alert and wide awake. You play a significant role in helping to ease the tension, thus allowing him to move into calm, deep sleep.

You can't make your child sleep, but you can set the stage

Every body also has a sleep/wake system that tells the brain when to be awake and when to be asleep. It's this body clock and the hormones and chemicals that go along with it that help the brain to know when to "switch" from alert to sleep. In this book, you'll discover that a good night's sleep begins in the morning, and that all day long you are making decisions that either help your child to fall asleep easily or innocently disrupt his sleep. Little decisions, like allowing your child a caffeine-laden soft drink at lunch, skipping a nap, or staying up late (just this time) can throw your child into "jet lag" and make it more difficult to sleep.

By becoming aware of the decisions you make each day, you can reduce the hassles and enjoy your child more—at least on most days. I'll be honest with you. Our goal is to increase the frequency of the "good days." I can't promise utopia, but I can assure you, it will be much better than it is now!

You may select strategies that fit your child

Every child is different. One child may shift into sleep within moments, while another takes at least forty-five minutes, and that's with your help. Some crave cuddles. Others squirm in your arms until you put them down and just let them sleep. Energy flows all day through certain children, and yet they conk out easily. Still others dash across the room at the mere suggestion of bedtime. Miss his "window" of sleep, and he's up for the next ninety minutes.

It's not a sign of failure if your child does not sleep like your sister-in-law's child. You're not doing something wrong when your child needs more help from you to wind down and ease into sleep. It's true that your child may need protection from the stimulation of the day. It's all right to discover what *your* child needs to sleep.

You can decide the best way for your family to sleep

Whether you choose to bring your child into your bed, have him snuggle up with a sibling, or sleep down the hall on his own is up to you. You won't find one method that fits all, in this book. Decisions regarding when and where you sleep, and whether you sleep with or without your children depend on your personal preferences and your family's values and culture. There really isn't one "right way" to sleep. So, what you will find instead are four simple goals:

1. Sensitive, responsive care for your children
2. Structure to help your children "fit" your family's way of sleeping
3. Sleep for *everyone*
4. Opportunities for intimacy for you as an adult

I'll help you use these goals as guidelines to find the best way for YOUR family to sleep.

You don't need to worry about bad habits; they can be changed

Inevitably, you will be surrounded by predictions of dire consequences, like: "If you start that habit, you'll be doing it forever." Yet sometimes your gut tells you that at this moment, your child needs something more. In this book, you will be given permission to adjust your strategies when life's stresses have piled up, putting your child chronically on "alert." Habits can be changed. You can meet your child's needs, help him sleep, and still gently nudge him back to more independence when he's ready. Consistently meeting your child's needs and getting sleep for yourself is most important. How you do it may change at different times, depending on the situation.

YOU WILL ENJOY THE DELIGHTS OF BEING WELL RESTED

As I have worked with families, helping them to find their way to sleep, the changes in the children's behavior and in the family's well-being have been thrilling. Each story is unique, but, like Jennifer's, often amazing. I first met two-year-old Jennifer when things were not going well. The whining would start the minute she woke up. She'd ask for something to eat, take a bite, and then throw the rest of the food on the floor. Immediately she'd demand something else—and scream hysterically if she didn't get it. In fact, every word out of her mouth was a command: "Get me this now!" Her parents nicknamed her Queen Jennifer.

The tantrums became regular occurrences, although the reasons for them varied. One day she went over the edge because her dad arrived home unexpectedly. A firm "no" from her mother triggered yet another forty-minute outburst. Her parents spent their days walking on eggshells as Jennifer rolled from one tantrum to another.

Between fits, Jennifer wreaked more havoc. Whenever her mother's back was turned, she would throw food, water, or toys in all directions. Even the family dog suffered from her behavior as she chased and hit him with all sorts of objects. Bedtime became a particular nightmare for Jennifer's parents. Their daughter would scream and struggle in their arms for ninety minutes or longer before finally falling asleep. Yet the peace was short-lived. Every night, Jennifer awoke, sobbing inconsolably.

Jennifer's parents tried punishment and letting her "cry it out" in hopes of turning her behavior around, but nothing worked. Finally, in utter desperation, they signed up for my class "Misbehaving or Missing Sleep?" They quickly learned that they'd been unwittingly making decisions that were undermining Jennifer's ability to fall asleep and stay asleep; decisions that they could change. By slowing down and making different choices, such as reducing the number of errands they were running, changing bathtime from nighttime to morning, and setting up a regular naptime, they were able to give Jennifer what she really needed: more sleep.

Within three weeks, Jennifer was sleeping thirteen hours instead of the nine she had been getting, and the tantrums disappeared. Today, Jennifer is well rested, and her behavior has completely changed. If she drops a toy, she will ask for help instead of collapsing into a screaming fit. When her mother says "no," she can be redirected easily to another activity. She also can be trusted to be in a room alone for a few minutes without causing chaos. Even the family dog has found peace.

The good behavior has spilled into the night as well. The former ninety-minute bedtime battle has been replaced by a twenty-minute routine, which includes a story, a back rub, and a sweet kiss good-night. Middle-of-the-night awakenings seldom occur, and when they do, they tend to be mere five-minute interruptions.

Jennifer's parents learned that no punishment, no threat, no discipline strategy could stop their daughter's challenging behaviors. What Jennifer needed was a simple good night's rest—not just once in a while, but every night. Jennifer had been literally shrieking for sleep.

When your child is well rested, the tantrums begin to disappear. A child who is not tired is calmer, more flexible, cooperative, attentive, and energetic. He gets along with others. You also benefit. Miraculously, as your child's hours of sleep increase, so, too, does your skill and effectiveness as a parent. Your own fatigue diminishes, and your confidence soars.

Ah, sleep, deep, sound, restorative sleep. It brings forth flexibility and patience, smiles and laughter. The difference between a child who is well rested and one who is missing sleep is a smile on his face—and on yours.

TWO

IS YOUR CHILD MISBEHAVING OR MISSING SLEEP?

"When the rhythm of sleep is thrown off, so, too, is the beat of life."

—Gary

Sunlight streamed through the skylights as I arranged the chairs around the table. This was my noon class. A one-hour, bring-your-lunch-and-learn event attended by sixteen parents, some running in during their lunch hour, others bringing their children along.

The topic on this day was "challenging behaviors." I began the session by asking the group: how many of you believe that your child has a sleep problem? Three parents raised their hands. John was not one of them. Six feet four inches tall, John towered over the rest of us. He favored a ballcap set backwards on his head and seemed surprised to find himself in a class. But his son's frequent tantrums made him willing to try just about anything. I noticed that his registration had been e-mailed at three o'clock a.m.

John slumped lower in his chair, his eyebrows knit closely together, as he asked: "What does all of this have to do with temper tantrums, talking back, fighting with siblings, not listening—the things I came to talk about?"

Lightheartedly, I pumped my fist in the air. "Thank you, John! I was hoping someone would ask that question." I turned to the group and asked for volunteers. Five parents who had children who were well rested, five who were moderately well rested, and five whose children were very tired. Then I asked them to describe their children's behavior.

The parents of well-rested children began. Their kids were happy and energetic, they said. Very focused, often involved in projects for extended periods of time, rarely fought with siblings, and got in and out of the car without a major fuss.

I turned to them. "So things are going pretty well?" They nodded in unison, except for Jenna, a tall, thin woman with wavy black hair.

"I have to admit, my daughter did get upset with me the other day."

"What happened?"

"I put vegetables in her macaroni."

The parents of tired children groaned, predicting what might come next. We waited expectantly for Jenna's response.

"She asked me to take them out."

"Well, what happened then?"

"I took them out."

"That's it?"

"Yes, and—well, she ate it."

The parents of the fatigued children moaned in disbelief. One struck the palm of her hand against her forehead. "I made my son macaroni, too. I put it in the wrong bowl. He screamed for thirty minutes and never did eat it, even though I switched it into the bowl he wanted."

The other parents of fatigued children could hardly contain themselves as they added their tales of woe. The temper tantrums are like stacked dominoes, one setting off the next one all day long. There is no "eye" in the hurricane of emotions. The complaining and fighting are constant, and you can't even think about asking them to pick something up—they react as though you had asked them to clean the entire house.

Suddenly I noticed Lynn, a quiet introvert, smiling sheepishly.

"Anything you'd like to add?" I asked.

"I think I'm forgetting what my daughter's 'normal' voice sounds like," she admitted. "She's always whining." And then, pausing a moment, she added, "I'm not certain she still has a normal voice." The others nodded in complete agreement.

The parents whose children were moderately well rested had remained silent, listening to the rather rowdy "fatigued" group. When it was finally their turn, they reported that their children weren't too bad—until late afternoon. Then things started falling apart.

MISSING SLEEP LEADS TO MISBEHAVIOR

The reality was stark. Well-rested children behave themselves—at least most of the time. They are more independent, helpful, and cooperative. Tired children get into trouble.

POWER STRUGGLES ARE ABOUT FEELINGS AND NEEDS

Behind every power struggle, every temper tantrum, every instance of misbehavior, there is a feeling and need—too often, that need is sleep. It's easy to miss the connection between the poor behavior and lack of sleep, thrown off by the whining, the "attitude," and tears. Our first thought isn't "Oh honey, are you tired?"

The exciting thing is that when your child's misbehavior is the result of missing sleep, you really can make a difference. It's you, the adult in your child's life, who can lay the foundation for sound sleep. So, how do you know if your child's challenging behavior is fueled by exhaustion? Let's take a look at the signs.

GETTING A PICTURE OF MISBEHAVIORS CAUSED BY MISSING SLEEP

As you review each category, think about your child's recent behavior. What response have you seen lately? How frequently is it appearing? (If you've read my first book, *Raising Your Spirited Child,* you may wonder how to know if your child is spirited or missing sleep. Spirited children, by their very nature, experience every emotion more powerfully. They must work harder to manage emotions. As a result, studies demonstrate, spirited children are especially vulnerable to the impact of sleep deprivation. Short on sleep, the passion and unpredictability of their emotions increases even more.)

Check all of the behaviors that apply to your child.

1. Emotions

How well is your child coping with frustration, dealing with surprises, and managing anger? To determine whether missing sleep is the culprit, look for the following reactions.

_____ Loses it over "little things"

_____ Is easily frustrated

_____ Becomes upset by changes in routine, or surprises

______ Is difficult to calm or comfort

______ Is irritable and cranky

______ Experiences frequent stomach- or headaches

______ Is unsatisfied; nothing is right, no matter what you offer

______ Is easily overwhelmed

______ Is anxious and resistant

Fatigue undercuts a child's resiliency and flexibility, leaving her more vulnerable to spillover tantrums. That's what John recalled. "It was a simple request to play downstairs that dropped Ethan to the floor. 'NO!' he shrieked and rocked his head back and forth, as he wailed, 'No! No! No!' I tried to pick him up to offer comfort, but he reeled in my arms, arched his back, and then collapsed in a heap on the floor."

Seven-year-old Thomas's reaction was not as demonstrative as Ethan's. Still, lack of sleep took a toll on his usually calm demeanor. "He sat at the kitchen table, attempting to complete his homework," his mom explained. "I heard a sniff, and turned to see tears running down his cheeks. 'What's the matter, honey?' I asked him. This was so unusual for him.

" 'My shoe fell. I didn't want it to fall,' he replied, tears slipping down his cheeks.

"He was weepy about everything," she continued. "When he didn't know the meaning of his spelling word, he cried. When I told him to wait for his snack—he bawled. Normally, those kinds of things wouldn't faze him. But he had soccer practices that ran late two nights this week, and it really wore him out."

Whether it's a ballistic explosion or a tear quietly slipping down your child's cheek, the issue is the same—emotional overload. Their emotions are more powerful, negative, and volatile when your child is exhausted. The only thing you can do at this point is attempt to calm your child down—which is not easy. The teachable moment will have to wait until she's gotten some sleep.

In contrast, well-rested children are on a more even keel—even the spirited ones. Transitions go more smoothly. Homework isn't as overwhelming. If you fail to offer them the right cup, or respond as quickly as they'd like, there's not a huge meltdown. They can wait and even talk it through. If something surprises or upsets them, they can be easily consoled. Little issues do not flare into bigger ones. They are happier—even first thing in the morning and late afternoon.

If you feel like you are just waiting for the next blowup, and your child is getting in trouble for tantrums, your first course of action to change this behavior is to help your child get more sleep. When you do, you'll discover that the "meltdowns" begin to mysteriously disappear.

2. The Body

Your child's body is also impacted as tension and fatigue rise. Movements become jerky, frenzied, and often impulsive.

How well is your child able to control her body? Are her movements smooth and energetic, or is your child "wired" and unable to stop? To determine if your child's behaviors may be caused by sleep deprivation, look for the following reactions:

_____ Clumsy, experiences frequent accidents, falls, and injuries

_____ Frenzied, hyperactivity

_____ Wild at bedtime, can't fall asleep—even when tired

_____ Hits, throws things, or shouts

_____ Has to be awakened in the morning

_____ Gets sick more frequently than the other kids

_____ Craves carbohydrates or sugar

_____ Is lethargic; can't seem to do what he is usually capable of doing

_____ Seems unable to stop from breaking the rules

Self-restraint takes energy. Tired, tense kids don't have the stamina to control themselves. Sheila saw the troubles lack of sleep caused for her son Sam.

"We spent the holiday weekend with extended family at a resort. Sam got to sleep at eleven fifteen one night and eleven forty-five the next. Contrary to my better judgment, we succumbed to family pressure for all of us to stay up for the parties. He did sleep later the next morning, but it wasn't enough. The next day, Sam completely self-destructed. He hit and screamed at people. He struck his father with a rubber baseball bat. He called people names, and I can't help believing that he intentionally spilled his brother's milk all over the kitchen floor."

While it might certainly have looked like a plot, Sam wasn't purposefully trying to be difficult. Exhaustion robbed him of energy. Restraining the surging forces within him took more self-control than he had available. He couldn't pull it back. His brain compensated by activating his arousal system. Like firing a rocket, it fueled a frenzy of motion that he couldn't stop.

The frenzy can also lead to fumbling and stumbling. Under the crushing weight of fatigue, children more frequently drop things, contributing liberally to the mess on the kitchen floor, or fall, adding to the collage of bruises on their legs, and maybe even another trip to the emergency room.

Fatigue can also take your child "down" by making him more susceptible to infection. Sleep scientists at the University of Chicago found that even in young, healthy people, a sleep debt of three or four hours over the course of a week affects the body's ability to process carbohydrates, manage stress, maintain a proper balance of hormones, and fight off infection.

As Dr. Karine Spiegel and Dr. Eve Van Cauter say, "Accumulated sleep debt is potentially as detrimental to health as poor nutrition or a sedentary lifestyle."

The difference in physical energy and self-restraint between a child who is rested and one who isn't is readily apparent. Eyes bright and twinkling, the well-rested child moves smoothly about the room. He can dress independently and even restrain himself from hitting his sister when she snatches his favorite bowl from the cupboard. Sleep comes easily and deeply.

If it feels like your child is frenzied and out of control, it's important to recognize that a little more sleep can be a "miracle cure." It's sleep that can help your child slow his body, honor the rules, be less accident-prone, stay healthy, and fall asleep more easily.

3. Attention, Focus, and Performance

When a child is overtired, his body compensates by releasing stress hormones into his system, which cause him to be more "alert." As a result, his brain automatically tells him to be aware of what is going on in his environment, in order to keep himself safe. The result is a child who has problems staying focused on important things, screening out trivial stimulation, and performing well.

How well is your child able to maintain focus? Does your child stay on task or become easily distracted? Is your child able to perform at peak level? To determine if your child is sleep-deprived, look for the following behaviors:

_____ Loses focus, wanders from one activity to another

_____ Seeks stimulation to keep going—annoys siblings or pets, wants to watch TV, especially in the late afternoon

_____ Needs your attention and help to stay on task

_____ Is forgetful

_____ Struggles to make decisions

_____ Doesn't listen

_____ Has difficulty performing at peak level or resists participating altogether

_____ Talks excessively

_____ Finds it difficult to work without disrupting others

"When a child is exhausted, he has to shift attention frequently and create enough commotion around him, to keep himself awake," says Dr. Ronald Chervin, from the University of Michigan.

If your child wanders aimlessly from one toy or activity to another, you may worry that he has an attention deficit disorder, which, indeed, is possible. But now you know that the first thing you need to check is whether or not he is getting enough sleep.

Lack of focus also affects performance. All too often we've been taught to ignore fatigue, or push through it, but today research confirms that without sleep we lose focus. Without focus, performance deteriorates. The relationship between sleep and peak performance has been so well documented that Duke University, recognizing the cost of sleep deprivation for its students, has eliminated eight o'clock a.m. classes. And we now know that if your child has the choice between studying another hour for an exam the next day or going to sleep—you'll want to send him to bed. He'll perform better.

Another benefit of sleep is greater independence. Children who are well rested are more likely to experience calm energy. In a state of calm energy, all systems are in balance, allowing them to stay engaged in the task at hand. They don't need your assistance or energy to keep them awake and on task.

If your child is struggling to stay focused and pay attention, it's important to recognize that sleep—or lack of it—may be the real culprit.

4. Social Situations

If it seems as though your child has lost all of his social skills and is constantly arguing with you or others, lack of sleep may be the cause.

Is your child able to get along with others, or does he frequently end up in arguments? To determine if your child's behaviors are caused by sleep deprivation, look for the following reactions:

______ Experiences hurt feelings easily

______ Has difficulty being patient

______ Suffers from separation anxiety

______ Is bossy and demanding

______ Loses it if told "no"

______ Has difficulty solving age-appropriate problems, or talking things through

______ Easily forgets the rules or wants to debate them

______ Is irritated by siblings and peers—especially in the late afternoons

______ Is not open to your guidance

It takes effort for young children to understand their own emotions, let alone those of their friends and siblings. By late afternoon, tired children have trouble remembering social rules, being empathetic and patient. In social interactions, these lapses lead to squabbles and even physical fights. Sarah found this to be very true.

"The neighborhood kids always play outside after school," she explained. "But that's when Jason has the most trouble. Last year, I had to supervise him. If his feelings were hurt, he'd hit someone. This year, he's eight. At least, he's not hitting, but he'll stomp inside angry, pouting that something isn't fair or someone won't play with him. Initially, I didn't know if he was tired or if it was just who he was. But then I began to realize that during the week, he was averaging eight hours of sleep a night and getting into fights. On the weekend, he got ten hours, and played beautifully all day long."

Curiously, the conflicts are not just limited to peers and siblings. You suffer, too, when your child is short on sleep. That's what Todd recognized.

"Ryan refuses to cooperate in the morning. He'll complain, 'I'm too tired. Why did you have to get me up so early? I hate this shirt'—mind you, he wears a uniform—'these pants bug me. I don't want a waffle. Why do you keep bossing me around?' "

If your child is arguing with playmates and with you, before you send him to "time out" or blame it on his temperament, try helping him to get more sleep.

Dr. Judith Owens says, "Sleep habits are not typically considered when kids are assessed for behavioral or learning issues. Absolutely every child who's being evaluated for academic, learning, behavioral problems, or ADHD should be screened for sleep issues."

PUTTING IT ALL TOGETHER

When children are exhausted, it's rare that they will tell you. Instead, if you watch carefully and understand what you are seeing, their behavior will show you. They'll moan and groan, become helpless or downright nasty. They may become "wild," running and hooting through the house, or down the aisles of the grocery store. However they express it, the message is the same. They're tired, and their behavior is deteriorating with every step. They need you to help them choose sleep.

Now go back through each of the four areas and total your responses. Mark your total on the scale below. How tired is your child?

TOTAL SCORE

0–5 WELL RESTED 6–15 TIRED 16–36 REALLY MISSING SLEEP

This is not a scientific test, or a research project. But if your child's score falls within the 16–36 range, you can assume that sleep deprivation is pushing your child's system out of balance. Go back through the lists and take note. Is your child especially vulnerable to losing focus and attention when he's tired? Do emotions overwhelm him? When he's exhausted, are social situations his downfall, or is it impulse control that challenges him the most?

Now, when you see these behaviors, you don't have to feel out of control or wonder what's wrong. It will be clear to you that your child needs more sleep. It's something that you do have an influence over. This awareness will provide you with a starting point for changing your child's most challenging behaviors. By choosing sleep, both you and your child will benefit.

If it feels like an overwhelming task, don't despair. Admittedly, making the necessary changes requires energy. Perhaps more energy than you feel you have right now. Your needs are very important, too, so, let's take a look and see if you also need more sleep.

THREE

Short-Tempered, Feeling Overwhelmed: ARE YOU MISSING SLEEP?

How your fatigue affects your child's behavior

"It never occurred to me that some of my daughter's misbehavior could have something to do with my own lack of sleep."

—Sarah, mother of two

Tall, athletic, with a slowly receding hairline, Paul pulled open the door to the family center. I smiled, happy to see him. He was a quiet, observant group member, whose clever sense of humor frequently left us delightfully flabbergasted and created a refreshingly playful ambiance in the group.

Baby Jacob was clutched in a football hold under his arm, and Katrina, his three-year-old daughter danced at his feet. "But why is the sky blue, Daddy?" she asked. I saw his jaw tighten. "How many times do I have to tell you, Katrina, I don't know. We'll have to look it up." The sharpness in his voice caught my attention. Something was up. This was a guy who might roll his eyes at his daughter's unending questions, but he answered them, or left her laughing with a witty quip.

Ignoring her father's tone, Katrina declared, "Let's do it now!"

"No!" Paul responded sternly. Now whining, Katrina complained, "But Daddy, you said we could look it up."

"Stop it, Katrina. Stop it right now or we're going home!" Paul threatened. He sighed deeply, and then glanced at me, a hint of desperation in his eyes. It was then I pointed to the board where I had written our discussion question for the class. I raised my eyebrow as though to ask, is this it? He read:

When you are tired, how does your response to your child's behavior change?

"Oh, you don't want to hear my answer," he replied, almost smiling.

I took Katrina's hand. "Let's go ask Teacher Stacey if she knows why the sky is blue." Katrina galloped ahead of me into the children's room. Paul sighed again—this time, in relief. Soon the other parents and their children began meandering in. They, too, stopped to comment about the "interesting" question.

It's not just the children who are impacted by lack of sleep. Sleep deprivation creeps up on you, too, wearing thin your patience, compelling you to sound like your parents, even though you promised yourself you never would, and losing it over things that on a good day would never take you down.

SOUND SLEEP ENHANCES RELATIONSHIPS AND CHANGES BEHAVIOR FOR EVERYONE

I once saw a T-shirt with the message printed on it: AIN'T NOBODY HAPPY, IF MOMMA AIN'T HAPPY.

Sleep researchers have now demonstrated that sleep is one of the most essential elements for making Momma—or Daddy—happy. William Dement, a renowned leader in the field, states: "The effect of sleep deprivation on mood in the average healthy person is a huge, invisible problem." Studies consistently show that lack of sleep increases irritability, volatility, and depression. Even minor irritations can feel overwhelming when you're tired.

Her deep brown eyes made even darker by the gray, smudged rings under them, Emily explained to the group, "It was the tone of her voice that brought the hair on the back of my neck to standing. I'd simply asked my four-year-old to go to the bathroom, but she turned up her chin and flatly refused to go unless I helped her. I was busy, and told her to just go. She wouldn't. Two minutes later, she was demanding to go NOW! Gritting my teeth, I helped her out of her leotards. She dashed to the bathroom, only to trip and bump her head. The scream was blood-curdling. I grabbed an ice pack—except I grabbed the wrong one. She shoved it away, insisting, not the pink one, the blue one. I grabbed the blue one. This, too, was shoved away. My third and fourth attempts to placate her were a dismal failure. She collapsed in the hallway, shrieking.

"I'm not going here, I thought in near panic. I'm not going to let her run our lives! I've got to set a limit somewhere—but then she was kicking the wall."

Emily exhaled slowly. "I know, we never would have gone there if I hadn't been so exhausted that day. But when I'm tired, my fuse is about a centimeter long, and she burns that up really quickly.

"The irony is that two days earlier, I had also asked her to use the bathroom, and she had refused. But that day, I wasn't feeling groggy or sluggish or just plain exhausted. I had the mental and physical energy to keep her on track and stop her from getting to that overload point in the first place. When she said no, I didn't react. Instead, I responded the way I'd learned in class. I asked her a question, to shift her brain into thinking. I asked her, 'Would you like me to count to fifty or one hundred while you give it a try?' She thought about it for a moment, asked me to count to one hundred, and then trotted down the hallway, focused on the numbers instead of fighting with me."

Power struggles require two people. Someone has to push and someone has to pull. Sometimes it's easier to focus on your children's behavior than your own, but relationships are a two-way street. On the "good days," when your child becomes upset, your response is different. It's much more likely that you will have the stamina to stop and creatively solve the issue. But when you're tired, weighed down by the tension of the day and lack of sleep, the tendency is to take the shortcut—the one that ends up costing you in the end. As a result, your ability to prevent the struggles and bring your child back into balance diminishes exponentially. As Paul said later that evening, "I've been reading books and talking to people about the importance of sleep for my kids, but I haven't considered my own needs."

It's easy to miss the link between your own lack of sleep and power struggles

Perhaps you know you're not getting enough sleep. The kids keep waking you up, or there just doesn't seem to be enough time in the day, so you're staying up to grab a few quiet moments or finish a project of your own. But what you might not be aware of is the toll that lack of sleep is taking on your mood, body, focus, relationships, and even your finesse as a parent.

So, let's take a look and help you identify the signs that indicate you are tumbling into the pit of sleep deprivation. Catching yourself allows you to stop the downward spiral for you AND your child at the first hint of trouble. It's this awareness that can help you bring your entire family back into equilibrium, and fill your days and nights with a sense of calm—instead of tension. You really can do it.

Are you missing sleep?

Stop for a moment and ask yourself—how am I feeling right now? How tense is my body? What is my energy level? If no answer comes, give yourself a few minutes. Perhaps you are so tired that you are no longer able to identify your feelings. But those "feelings" are your cues—the signals that tell you what you need at that moment to keep your body in balance. The ability to be aware of your own tension and energy levels benefits your children, because the more practice you have picking up your own sensations, the easier it will be to sensitively respond to theirs.

When you are missing sleep, research demonstrates that it's easier to slip into a foul mood. As a result, negative emotions, such as irritability and anger, occur frequently and become more challenging to manage. Review the following list. Check all of the items that apply to you when the experience of feeling energetic and calm throughout the day no longer exists in your recent memory.

1. Emotions

How well are you coping with frustration, dealing with surprises ,and managing anger? When you are missing sleep, you will find yourself more likely to:

______ Be short-tempered, easily "set off" by the kids

______ Feel irritable and cranky, nothing is much fun

______ Burst into tears

______ Become frustrated easily

______ Find it difficult to alter plans or deal with surprises

_____ Become controlling and demanding

_____ Feel overwhelmed, anxious, or jittery

_____ Experience head- or stomachaches

_____ Feel guilty that you just don't have the energy for the kids

Terri grimaced as we went through the checklist in class. "I can't stand the noise and mess when I haven't had much sleep," she said. "The racket drives me crazy. I've got four kids. There are lots of opportunities for me to go ballistic. My poor children, they don't stand a chance of pleasing a mom in my state." She shook her head as others agreed.

Jamie was the parent who always chose the chair at the corner of the table, the one that gave her the most space. When I turned to her, her eyes were shiny with tears. "I feel so overwhelmed," she said softly. "My daughter comes close to me and touches me. I smile, but I feel like crying."

If you, too, find yourself feeling guilty knowing that you are more sensitive or reactive, and weighed down by everyday routines, sleep—or lack of it—may be the mystery culprit.

2. Body

Your body also pays a price for sleep deprivation. When you are missing sleep, your energy level drops, and tension and fatigue rise. Coordination deteriorates, and it becomes more difficult to accomplish tasks or even move.

How well are you able to control your body and impulses? How is your energy level? When you are missing sleep, you will find yourself more likely to:

_____ Feel frenzied, unable to stop yourself

_____ Have difficulty falling asleep—even though you're tired

_____ Hit, yell, or throw things

_____ Wake up in the morning to an alarm going off or a child waking you, rather than on your own

_____ Become ill more frequently

_____ Crave carbohydrates, sugar, and/or caffeine

_____ Drop things, stub your toe, turn your ankle, or stumble

_____ Feel sluggish, heavy, unable to make a meal, pick things up, respond to a child

_____ Experience overwhelming sleepiness at certain points in the day

"I just want to hit him," Tim confessed, as we went down the list, then quickly added: "I don't."

"Good!" I responded just as fast.

"I know," he said. "Spanking isn't where I want to go. But when he won't even give me five minutes to change my clothes, or is standing outside the bathroom door, asking, 'Daddy, what about this?' or 'Daddy, can we do this?,' I think of Jackie Gleason and I want to send him to the moon!"

And it was Natasha who pointed to a check mark next to craving sweets on her handout. "I'm trying to eat a healthy diet," she said. "Last Wednesday, I'd been running all day long. By the time I got home, I craved a cookie. My mother-in-law was visiting. She had cooked a delicious meal for us. I sat down to nurse the baby and saw the entire meal on the table. I asked my husband to bring me a cookie. He said, 'What about a salad or some ham?' I nearly shrieked, 'A cookie. I NEED a cookie!' "

If you, too, find yourself shouting, throwing things, or yearning to let loose and hit when you know this is not your typical response, sleep deprivation may be stretching the limits of your self-control. And if your shins are bruised from bumping into end tables, or tripping as you head for the cupboard searching for sweets—it may be time to think about your own need for sleep.

3. Focus, attention, and performance

When you feel energetic and calm, you are relaxed and can stay focused. But when you are missing sleep, your body goes on "alert," making it much more difficult to pay attention and to perform well.

How well are you able to concentrate and to perform? Are you easily distracted? When you are missing sleep, you will find yourself more likely to:

_____ Feel as though you are in a fog

_____ Mix up words

_____ Forget things

_____ Make a list and then lose it

_____ Perform poorly—especially on things that require quick thinking or action

_____ Miss "cues" from your children and others

_____ Miss your exit on the freeway

_____ Have difficulty making decisions or thinking things through

_____ Find it impossible to be creative

Amy began to chuckle. "Thank you," she said. "I thought I might be developing a disability. But now I know that 'forgetting' is part of the price I'm paying for mothering twins. Yesterday, I went to the discount store. I parked the car and walked in with the babies in their stroller. When I came out later, I couldn't remember where I had parked. I wandered the lot for twenty minutes, on the verge of tears, before I finally found it."

Kelly grabbed Amy's arm and almost squealed. "Yesterday I put the baby in the car and drove away, leaving the garage door wide open. When I came home, I saw it and yelled at my husband for doing it. The poor man had been at work all day."

If you find that the kids are out of control and you realize you never saw or heard it coming—you may be too tired to be picking up the cues. And if you find yourself wandering on the wrong freeway or leaving the sack lunch you just prepared on the counter—it's time to talk about sleep.

4. Social Interactions

Tension and fatigue take over when you are missing sleep, leaving you with little tolerance for others. It's when you are tired that you have the hardest time being the kind of parent you want to be.

How well are you able to manage social interactions with your children and others? Are you able to get along, or frequently disagreeing? When you are tired, you will find yourself more likely to:

______ Argue with your partner and your children

______ Take your child's behavior more personally

______ Pick battles you would never pick on a "good day"

______ Put more pressure on your children

______ Find the sibling squabbles intolerable

______ Demand that things be done NOW!

______ Be more easily hurt by the comments of others

______ Be less flexible

______ Allow the "tone" to creep into your voice

"Forget the empathy. That's the first thing to go for me," Christine quipped. "Last week, I flew across the country with my three kids, to visit a friend. I got up at five o'clock a.m. to get myself ready and then woke them. Needless to say, we were all a mess. That evening, Jack kept coming out of the bedroom, saying he couldn't sleep. I knew it was a new environment, and that he was having trouble with the strange bed. He wanted a drink. He needed a different stuffed animal. But that night, it was all overwhelming. I can't do this, I thought, and ended up threatening that if he didn't stay in bed, we wouldn't go to the amusement park the next day. The threat was completely unrelated to the issue, and as I heard myself utter it, I couldn't help thinking, who is that speaking?"

If your buttons are getting pushed more easily, or you find yourself sounding like your ten-year-old, it's important to recognize that sound sleep is an essential element for bringing the laughter and good times back into your home.

Now go back through this chapter and count how many items you checked. Add them together to determine your score.

OVERALL SCORE

0–5 WELL RESTED 6–15 MODERATELY TIRED 16–36 NEEDING SLEEP

Your sleep needs matter

This is not a scientific test. It's merely a guide for you, to help you recognize the feelings and behaviors you experience when your body is drained and needing sleep. Is sleep deprivation playing havoc with your relationships? Is there a certain area where you are more vulnerable? Are you more tired than you ever imagined? Are the effects of fatigue creeping into every word you utter and every action you take?

"I checked every single item," Jamie sighed as we finished the list in class. "I thought I was stressed. I never couched it as sleep. I can actually do something about it." Her voice rose in excitement. "People say parents don't get enough sleep, but they never really talk about what goes along with it. I've certainly been more emotional, snapping at the kids, but I didn't realize it could be as simple as sleep. It makes me feel so much better. It's not this inherent ugly part of me, or my daughter. It's so much more reasonable and controllable. It gives me hope!"

Jamie was not the only one who was comforted. The relief of the entire group was palpable. They weren't bad parents. Their children were not out to get them—they were simply tired.

Becoming a "sleep manager"

Once you recognize the signs that you and your child are missing sleep, you have a guide—a map that will lead you and your child out of the conflicts flamed by fatigue. It really is possible to make it better, and the next week, Jamie confirmed that she had already put her newfound awareness to work.

"I left Dani with a sitter yesterday afternoon," she said. "A storm hit while I was gone, and the basement flooded. I rushed home. My mom came to help, too. Dani missed her nap completely. There was a lot of energy in the house that afternoon. By dinner, she was talking nonstop, playing with her food and refusing to eat. It was driving me crazy. Before, I would have gotten upset with her, but this time, I recognized she was tired and so was I. I asked my husband to put her in the tub. Usually, I sit with her, but I asked him to do it, because I knew I was too exhausted. I couldn't tolerate the chattering anymore. I needed quiet. I did the dishes. He bathed her, and by seven-thirty, the kitchen was cleaned up, and she was sound asleep. No power struggles, no tantrums—just peace. Now I understand what's happening. It's something that I have an influence over. I don't feel so inept or scared anymore."

Sleep is like sunshine. It brings energy to your day, makes you happy to be alive, and eager to face whatever life tosses your way. Now that you recognize your entire family's need for sleep, the question, of course, is how do you get your child to sleep?

LEARNING ABOUT SLEEP FROM A NEW POINT OF VIEW

FOUR

HOW DO YOU GET YOUR CHILD TO SLEEP?

Sorting through all of the advice

"It's true that children can cry themselves to sleep, but I still wonder, do they fall in exhaustion or in despair?"

—Robert, father of three

I was troubled. The previous week, the parents in my group had protested, declaring that they'd tried to get their kids to sleep using all of the advice they'd been given by professionals, friends, and family members, yet it hadn't worked. I asked them what advice they'd received, and together we brainstormed a list. It looked like this:

They need to learn to soothe themselves

Don't let them in your bed

Insist that they sleep in their own room

Don't feed them during the night

Put up a gate to keep them in their room

Let them cry

Put your foot down—take control

Don't let them manipulate you

Shut the door

Don't hold, rock, or lie down with them

Don't nurse them to sleep

Don't stay more than two minutes with them

They need to learn to be independent

There had even been a pamphlet that Amy had picked up, which recommended that if your child cried so hard that he vomited, you should not clean it up until the next morning, or you would be reinforcing the behavior.

After class, I pondered that list. It troubled me.

That same week, my father fell and broke his hip. Once six feet two inches tall, and two hundred fifty pounds, he had been a strong, muscular man who in his prime could throw open the fifteen-foot machine shed door with one arm. Now Parkinson's disease imprisoned his sound mind in a body unable to roll over in bed. His latest fall and the resulting broken hip finally forced my mother, sisters, and me to seek help caring for him. As we visited different facilities and interviewed staff, I wasn't as concerned about his daytime care as I was about the nights. I knew that someone from his family or cluster of friends would be with him during the day, helping to care for him, but at night he'd be alone. I agonized, imagining him lying alone in the dark, unable to roll over, get his own drink, or take himself to the bathroom. If he experienced pain, or simply needed someone to be with him, would he be heard? The Parkinson's disease had weakened his throat and chest muscles, leading him to often aspirate his food. He frequently choked and vomited. Would anyone respond to his calls for help? And so, at each stop, I asked: "What is your nighttime policy?"

I shared the dilemma of my family's decision with my class the following week and decided to read to them the policy for facility A, to see what they thought. It read like this:

> Nighttime Policy A
>
> In this facility, we encourage patients to be independent. Each individual has his or her own room and is expected to stay there all night. If the patients are mobile and try to get out, we put up a gate to stop them. Ultimately, it works. If they scream loudly, we close their door, to prevent them from awakening the others. We don't let them manipulate us. If they get so upset that they vomit, we'll clean it up in the morning. They'll learn not to do it next time. We run a tight ship here, clearly letting patients know that we are in control. There's no coddling. We can't be running to check on them every time they want something. We've got other things to do. We're not going to sit and hold their hands. They've

> got to learn to sleep through the night, or at least remain quiet if they don't. And if they're hungry or wet, our staff is scheduled to begin rounds at six o'clock a.m. It's important that they learn to wait.

I turned to them and asked, "What do you think? Should we place my dad there?"

Jenna was incredulous. "That's just cruel! You wouldn't!"

His voice tight, eyes sparking anger, Paul snapped: "That's inhumane."

"Heartbreaking," Kelly almost wailed. "They don't need to be taught independence. If they were independent, they wouldn't be there."

Lynn had sat, silently listening. Suddenly, she thrust her body forward, jabbing a finger at me. "Is that true?" she demanded. "If that's true, you need to call the state and shut that place down!"

At this point, I had to admit that while the search for a care facility was true, the policy was not. I explained quickly how the advice they'd received had gnawed at me, especially when I thought of my father, who now, like a young child, was completely dependent on others for his care. I pointed out to them how I'd written the "advice" they'd received into the policy. And as we reviewed it once more, I wondered out loud, why, when we hear this advice given for the elderly, do we find it offensive, even abusive, and yet we accept it for children? The silence in the room was heavy, as each person was filled with their own thoughts.

A NEW POINT OF VIEW

I asked them then, what advice do you wish you would receive? It was Sheila who said, "I'm tired of feeling guilty when I do what I know helps Sam to sleep. I want people to say, 'Every child is unique. It's all right to comfort him and listen to what he needs. One strategy doesn't have to fit all.' "

"Watching my oldest, I've come to realize that nothing lasts forever," Terri offered. "I want to say, 'Do what works for your family. There isn't one right way to sleep.' And if you decide you don't like how you're sleeping anymore, you can change it."

Jamie's voice was soft, almost a whisper, "Listen to your heart."

And then she told us that, despite admonishments to stop, she had rocked her son every night for seven years—until he died of complications from a neurological disease.

As the group members talked, I made a list:

Every child is unique, as is every situation

It's all right to offer comfort

Do what works for you and your family

There's no one right way to sleep

One strategy does not fit all

It's all right to listen to what your child needs

Sometimes it takes time to make changes, but you can

Listen to your heart

I took the "wish list" home with me and, from it, created a nighttime policy for facility B. The next week, I read it to the class.

> Nighttime Policy B
>
> Every patient is unique. We never forget that. The sounds and lights here are usually very different from what they're used to at home. Some settle right in, but others take several weeks, even months to feel comfortable.
>
> We check on them frequently. Some just need to know we are here. Others want us to stop, sit with them, and hold their hands for a few minutes, or rub their backs. They need that touch, closeness. You know many are used to sleeping with someone else, and they miss that sensation of warmth and touch. We add an extra pillow or a rolled blanket if we're not worried about them suffocating or getting caught; it gives them a little extra comfort against their body.
>
> We keep a regular routine, so they know what to expect. And if they call, we will come. It's important for them to know. We find they rest better, and it improves the quality of their lives.

> There's no one right way that we do it here, but what you can count on is that every individual will get what he or she needs. We try to never forget to listen to our hearts. That's why we're in this business, you know.

Paul sighed. "It feels so much more respectful."

And Jamie needed to know, "Does the care facility you found for your dad have a policy like that?"

Fortunately, I was able to report that I had used it as a guide, and the facility we had found for my father did offer responsive and warm care. He was safe and attended to—even in the middle of the night.

WHAT THE RESEARCH TELLS US

In order to sleep well, your child has to feel safe. It's not merely an issue of philosophy but a reflection of physiology. Dr. Megan Gunnar from the University of Minnesota states, "Sensitive, responsive care blocks the stress reaction. If a child does not feel safe, she will get more anxious and stressed." And Dr. Ron Dahl, University of Pittsburgh Medical School explains, "The pendulum of arousal, moving between sleep and vigilance, is largely influenced by a relative sense of safety versus threat."

Anything that upsets your child's sense of well-being will raise her arousal and pull her system in the direction opposite of sleep. That's why it is important to look at the advice you have been given. Scrutinize it carefully and determine whether the recommended strategies create a sense of security that calms your child's body, thus gently nudging her toward sleep, or leave her feeling anxious and insecure, pushing her away.

HOW WE GOT HERE

Unfortunately, strategies recommended for children are not presented in a "policy" handbook that allows us to examine them carefully. It's rare that we even think of considering whether a particular strategy slows a child's heart rate, soothing her body, or escalates her heart rate, activating her stress system. Often, we simply fall into our methods without really thinking about how we got there. It's our emotions that can lead us to try strategies we never expected to use. Emotions like:

Hopelessness

"When Hannah didn't fall asleep, I saw it as a punishment of me," Emily offered. "I swear, on some days, I thought she was doing it to spite me. I'd get caught up in controlling something, anything. If she wanted the hallway light on, I would insist that she turn it off. Really, what difference does it make if the light is on in the hallway? Yet I'd fight about it for two hours, and then, because I was exhausted, I'd give in. What kind of parent can't even get her child to bed?"

Exhaustion

"I can't believe what comes out of my mouth when I'm tired and I just need my son to stop talking and go to sleep," Sheila said. "The other day, I told Sam that every time he came out of his room, I was going to take away another one of his stuffed animals. He loves his stuffed animals, he goes to sleep snuggling them, but yesterday I was taking away every one of them. It was all I could think of to do."

Fear

That's what happened to Sarah. "When I'm with my family, I find myself having a much shorter fuse with the kids. If they're having trouble settling down and I go in to read to them, I can hear my parents in my head. I know, they're thinking this is ridiculous. You're supposed to take them by the hand, walk them to the bedroom, and walk out. If they cry, you must be doing something wrong. Even when Jason was a baby, my mom would get upset with me. He was terribly colicky, so, I would hold him. It didn't seem to help much, but at least it made me feel like I was doing something. My mother would call and say to me, 'Are you holding that baby again? If you're holding him, you're going to spoil him.' I know she was worried about me starting a 'bad habit,' because that's what she was taught. I just wish she could have understood what it was like."

Helplessness

"The bedtime battles and night waking had been going on for weeks, no, maybe months," Lynn lamented. "We were desperate for sleep. Finally, when someone suggested we just let her cry, we decided to do it. It didn't feel right, but it had worked for my sister-in-law, and nothing else had worked. We tried for three days and then gave it up."

It's very likely that your heart has fought the use of strategies that leave your child feeling tense and threatened, but you might not have known what else you could do. Or you may have felt trapped, reluctant to ignore the warnings of others, or pressured to use strategies that so many others have. And it is true. Children may cry as they go to sleep. The key is in knowing the differences in the cries.

SENSITIVE, RESPONSIVE CARE MEANS LISTENING TO THE "CUES"

Lay one child down, and he may cry for a few minutes. A mad cry, as though to say, "This is hard work! I don't like it. I don't want to rest," but in less than five minutes, he falls blissfully asleep. As his parent, you realize that a bit of fussing was just what he needed to release the tension from his body and that he will now sleep well.

Lay another child down, and he screams as though he's pleading, "Help me, please help me. I can't stop!" And, indeed, he can't. His heart racing, eyes wild, hair mussed, he is unable to bring his body back into balance and calm himself. If left unattended, he will cry for hours, overwhelmed by the rush of stress hormones in his body. He cannot stop until someone helps him, not because he's trying to be manipulative but because of the tension and level of arousal in his body. Or, if he does finally "crash," as a parent, you are left wondering, as Robert did in class, does he fall in exhaustion or in despair?

When you practice sensitive care, you recognize the difference between the cries of these two children, and respond to each appropriately. If, however, you allow the advice of others, no matter how well intentioned, to stop you from listening to your child's cues and to your own heart's reaction, you lose your rudder, that deep sense of direction that tells you what your child needs and how to respond. Children can learn to fall asleep and to stay asleep with strategies that gently and respectfully get them there. You don't have to leave them screaming in the night.

In his book *The Heart of Parenting,* John Gottman writes, "Children who are responded to actually cry less, are able to wait longer, and are more flexible. They're also more open to your guidance."

Beginning today, you can choose to stop using strategies that disconnect you from your child and leave him too agitated to sleep. Instead, you can choose to "connect," listening carefully to what he needs. This does not mean that you will not get any sleep, or that you'll spend the rest of your life hopping from one child's bed to the next. Responding to your child in a sensitive way includes picking up the "cues" AND laying out a plan of action that "fits" him and teaches him how to sleep in your family.

SENSITIVE, RESPONSIVE CARE INCLUDES STRUCTURING FOR SUCCESS

During summer vacations, my daughter Kristina teaches at a golf camp. The children, ages 7–12, are there all day, and so the counselors are challenged to come up with games and contests to keep the kids involved. One of their favorites is a chipping game. The goal is to come within a foot of the hole. If it does, the counselor "owes" five push-ups. If the ball lands more than a foot from the hole, the camper "owes" a push-up. If it falls more than three feet from the hole, the camper is down for five push-ups, and if it stops more than ten feet from the hole, the camper is doomed to do ten.

On this particular day, the contest included five holes. Daniel, who was as round as he was tall, played with little enthusiasm. By the end of the game, he was down for fifty push-ups. But Daniel couldn't do them. It wasn't as though the other kids were pumping them out, either, but after Kristina showed the group how to do the "easy" version, from their knees, they each finished their set. Daniel collapsed in exhaustion after twenty.

"I couldn't let him just quit. I knew if he did, he wouldn't feel good about himself," Kristina told me later.

So she dropped to her knees on the ground next to him and

wrapped her arms around him. Every time he dipped to the ground, she gently pulled him back up. He hardly noticed her assistance. Thirty times she did it, until he sat up, red-faced and grinning. He'd done it. He'd finished all fifty. Later that day, a newfound confidence in his swing, he chipped like a pro.

Sometimes, when your kids are learning new skills, you have to drop to your knees, wrap your arms around them, and coach them through it, believing, as you do it, that they can be successful. The key is that you are working together. They, too, are putting in effort.

PUTTING IT ALL TOGETHER

Studies completed at the University of Minnesota by Ruth Thomas and Betty Cooke demonstrate that the most effective parents respond sensitively to the cues of their children AND create a structure for success. When it comes to helping your kids to sleep, that means breaking into steps the skills of calming oneself, shifting into sleep, and returning to sleep in the middle of the night. It also requires practicing every step and gently nudging your child toward achieving each skill. For some children, that "ladder for success" you help them create will be a mere stepstool, for others, it will require a sturdy twelve-footer, and others still will need you to add an extension, but by working together, you will get there. And you will do it in a way that "fits" your child and your family.

If a little voice inside you worries that you may be coddling your child when you help her learn how to soothe and calm herself in order to prepare her body for sleep, just remember: When your child first learned to stand up or to ride her bicycle, you put out a hand to steady her. You supported her as she practiced. Like these skills, learning to fall asleep and stay asleep sometimes needs a steadying presence. You are it.

TAKING A LOOK AT YOUR STRATEGIES

Stop and reflect. How are you approaching sleep now? Does your nighttime routine match the kind of nurturing care you are providing your child during the day, or are you doing things at night that you would never consider trying during the day? If someone asked you to post your "nighttime policy" at your door or on the Internet, would another family want to send their child to you for care? If you were a child, would you want to sleep in your home? Let's take a look.

It's ten o'clock p.m., a school night, and your seven-year-old is not asleep. Not only is he not sleeping, but he is also fighting with you, refusing to stay in his room or even lie down. It's only a matter of moments before his antics will wake his two-year-old brother, whom you just got to sleep.

You're exhausted. After work, you rushed home to eat dinner, and then dashed out the door again, piling the whole family into the car for the school musical. It was nine o'clock p.m., well past the kids' normal bedtime, when you arrived home. The dishes from dinner and breakfast are piled in the sink. The mail has not been sorted, and you're not sure if there is clean underwear for everyone to wear tomorrow. You attempt to shorten the bedtime routine. That's when the fight begins. The question is, what do you do next?

I hand out index cards to each class member, asking them to write down potential responses. They're invited to include strategies they've used themselves and others, recommended to them but never tried. "Do we have to admit we used them?" Christine groans, shifting uneasily in her chair. The others look up to hear the answer. I promise them they can say as little or as much as they like. Completed cards are tossed onto the table and shuffled.

Christine gingerly reaches for a card and reads it: "Threaten him. If he's not asleep in five minutes, he loses video games for the week." She lays the card back on the table with a smile. "I'm not the only one," she says. John raises a finger, to confess it's his, but at the same moment, Robert acknowledges it's his. They laugh, realizing there are three cards on the table that read "threat."

Robert flips over another. "Lock his door," he reads. "Ah, someone else has tried that one. We did it when our son was two. Everyone told us to do it. I remember standing downstairs in the kitchen covering my ears, because I couldn't bear to hear his screams. Ultimately, he did stop, but I think it took hours."

"My parents didn't lock my door," Jamie replied softly. "They locked theirs. So I slept all night on their threshold with my teddy bear. I still need him to fall asleep."

Without bothering to pick up a card, Lynn offered: "By ten o'clock, I'm too tired to fight any battles. I just want to sleep. Can't you just put him into bed with you or lie down with him for a few minutes? Obviously, he's overtired and so are you. You might as well snuggle up and go to sleep." Paul agreed. "I jotted down, go back through the routine. He must need it to wind down."

As they talked, I made a list, filling an entire page:

Threaten to take away video games

Threaten to turn off the lights

Threaten to reduce the number of bedtime books you'll read if he doesn't hurry

Lock his door

Lock your door

Let him sleep with you

Lie down with him

Rub his back

Go back through the entire bedtime routine

Give him a massage

Read to him

Let him snuggle

Sit with him

Now, which one do you choose? How do you respond to your child? Do your actions make him feel threatened, thus escalating his heart and pulse rate, or do they help him feel calm and connected, so that he can sleep?

LISTEN TO YOUR HEART

Literally, the answer lies in your heart. Imagine for a moment that you are that seven-year-old boy. You have been in school all day, then were rushed through dinner and swept into the car and onto a stage to perform in front of hundreds of your peers and their parents. It's two hours past your normal bedtime. Your body is jittery, images flash through your brain, and you're exhausted. You want to sleep, but you can't. Now let's look at each potential strategy. Ask yourself, does it make you feel anxious and threatened, causing your heart to race, and thus driving you away from sleep? Or does it ease the tension from your body, slowing your heart rate and allowing you to move into sleep? You will know that you are providing sensitive, responsive care when you feel connected and calm rather than disconnected and tense.

Strategy	**Body's response**
Threats to take things away	Disconnected and tense
Rub your back	Calm and connected
Lock the door from the outside	Disconnected and tense
Sit or lie down with you	Calm and connected
Yell	Disconnected and tense

NOW IT'S YOUR TURN

Grab a piece of paper and list ten methods you've used to get your child to sleep and to return to sleep when he's awoken in the night. Now imagine you are your child. If you were lying awake in the night, would these words, actions, tone of voice, or body posture slow your heart rate and help you relax, or would you find your heart racing?

Remember, before locking your child's door you can choose to stop and think what it would feel like to be in his shoes—on the other side of the door. When he is unable to calm himself and you are inclined to "let him cry," consider how you would feel if your partner ignored your distress. It's those decisions that make the difference between connecting or disconnecting with your child.

A VISION OF WHERE YOU WANT TO GO IN THE LONG RUN CAN KEEP YOU ON TRACK

An essential question to ask yourself today is "What kind of relationship do I want to have with my child in the long run?" Do you want to know that he will listen and respond to you? Do you want a teenager who comes in after an evening out with friends and sprawls on your bed or sits in the rocker next to you, to tell you about it?

You are building that relationship every day, word by word and action by action, as you interact with your child and teach him new skills—even the skill of sleeping. It helps to stay on course if you think about building a relationship that fosters independence yet encourages a willingness to listen to one another and work together. It is this type of relationship that builds a sense of security and trust within your family.

Sarah shifted in her chair, distress darkening her eyes. "But what am I supposed to say to my family when they disagree with me or offer advice that I'm not comfortable using?"

RESPONDING TO THOSE WITH A DIFFERENT "VISION"

It's always important to remember that the advice others offer is well intentioned. It doesn't mean, however, that it "fits" your child or encourages the kind of sensitive, responsive care that calms him and keeps him working with you. You can choose what advice to take to heart and yet respond respectfully to that which doesn't fit. In class, we created a list of responses for Sarah. You might find it useful, too. It looked like this:

"That's an interesting idea."

"I'll think about that."

"I've read that, too."

"I'll get back to you on that one."

"I've heard that mentioned before."

"Yes, that does seem to work for some kids."

"Tell me about your experience."

If you're feeling really brave, you might even say, "What I really would like you to do is to check in with me. Ask me how things are going and reassure me that I will figure out what works for my child and my family."

CELEBRATING SUCCESSES

Two weeks later, Jamie beamed, eager to share her success. "My daughter used to wake up every morning at four o'clock a.m.," she began. "She wanted me to lie with her, or to let her in my bed. I can't sleep with her, so I had refused. The fight frequently lasted at least an hour and a half, and left both of us frustrated and exhausted.

"I still can't sleep with her, but I realized I wasn't listening to her. She needs someone to be near, but I also need my space to sleep. She's old enough to understand that, so we made a deal. If she is quiet, she can snooze in a sleeping bag in her little brother's room.

"Now, if she wakes at four o'clock a.m., she comes to me and I take her to her brother's room. If I stay calm and just take her there and help her get into her sleeping bag, she's back asleep in four minutes—and so am I. Now we're working on getting her into the sleeping bag without waking Mom! The funny thing is that once I started listening to her, she started listening to me."

Kim nodded in agreement. "I've noticed that, too. Michael is wired differently. Everything is more with him. He's higher, louder, and more intense. He's always been that way. It would take hours to get him down at night. I didn't recognize how 'wired' and anxious he felt. I thought he was just being difficult. Now I understand that he needs more help settling down. Together, we created a plan.

"We explained to him that it was important for him to sleep in his own bed because Mom and Dad needed their sleep, too. He didn't like it. He asked us if we would stay with him. We said yes, but we also gave him an idea of what we expected to ultimately happen. We clarified that we would always read to him and tuck him in, but that someday soon he'd be ready to fall asleep on his own.

"The biggest challenge was staying calm. If he sat up, we gave him a few minutes, then stroked his forehead, and gently laid him down again. We never forced him. We just kept working with him. It did take several weeks, sitting with him, holding his hand, and then gradually introducing the idea that we were going to step out but would check right back. We did, too. We came back within fifteen or twenty seconds at first. It was important that he knew he could trust us. The last few nights our routine has included reading, stroking his forehead a few minutes, giving him a kiss, and, ten minutes later, he has been asleep."

Robert laughed. "We've gone in the opposite direction. During the daytime, we would never leave Alexia crying alone in a room, yet we were doing it at night. It must have been so confusing to her. I guess we were so worried about following the advice we'd been given that we stopped listening to her. Anyway, we've got a king-sized bed, so we decided to just let her crawl in with us. And now, because she knows that we will respond calmly, she's more peaceful. She's not even waking up as frequently anymore."

I listened intently, delighted by their successes, the unique solutions for each family, and the joy they had found working together with their children. Most of all, I enjoyed seeing the gleam in Sarah's eye when I turned to her.

"I talked to my mom! I told her I wanted her to be proud of how I worked with my kids. I let her know I appreciated it when she called, and then I asked her if she would please just remind me that I could figure it out. She listened. She really listened. It was wonderful, and that's when I realized that my daughter needs the same thing from me."

START WITH LITTLE STEPS

Bedtime does not have to be a time to "disconnect." Sensitive, responsive care allows your child to feel calm and secure as he learns the skills he needs to live in your family, including how to sleep. It's the little things that can make a difference. When you listen to your child's cues, you are helping his body to shift from awake to sleep. When you break the skills necessary for sound sleep into simple steps and practice them together, you are laying the foundation for sound sleep AND a strong relationship. When you are old and vulnerable, it is this relationship that will allow you to trust your child to ensure that you, too, are responded to—even in the middle of the night.

SENSITIVE, RESPONSIVE CARE REALLY DOES LEAD TO SOUND SLEEP

You may be in a bit of a quandary right now, relieved to know that you really can listen to your heart yet worried that to do so means your child will never fall into a pattern that fits your family, or that you will be reinforcing poor sleep habits. Let me show you how sensitive care helps your child to physiologically switch to sleep.

FIVE

WHY KIDS DO OR DO NOT SLEEP

It's a matter of physiology, not a power play

"The hardest part of changing my child's behavior is learning to change mine first."

—Sheila, mother of one

I once conducted a workshop in a small rural town. The audience was seated at round lunch tables in the elementary-school cafeteria. Three people sat at the table directly in my line of sight, at the back of the room. When I started to speak, one of them began talking—not whispering, mind you, but talking loudly. I glanced in her direction, valiantly attempting to communicate via body language that the presentation had begun. She kept talking. I noticed people around her turn and look, disturbed by the conversation. She ignored them and continued. I found myself growing more irritated. I tried to catch her eye. It didn't work. My thoughts raced as my pulse rate escalated. Should I say something? I thought. I didn't want to embarrass her, but I also didn't want her to continue disrupting the others around her. I found myself sputtering as my blood pressure rose. Suddenly, I faltered, losing my train of thought. I gave the group directions for an exercise and stepped over to my hostess. Turning off my microphone, I whispered to her: "Do you have any idea why that woman in the back of the room is talking?"

She peered over my shoulder. I noticed her struggling to contain a grin as she turned back to me. "That's our Spanish interpreter. She's interpreting for the parents at her table."

"Oooooh," I replied, now laughing at myself and thanking my lucky stars that at least I hadn't embarrassed her or myself by saying something.

I called the group back from their activity and casually slipped into the presentation that our workshop was being translated into Spanish. Magically, all the annoyed glances being directed at the woman stopped, and the audience members once again focused on the presentation. But even more interesting to me was my own reaction. Once I understood what she was doing and why she was doing it, her behavior no longer upset me. I didn't even notice as she continued to talk throughout the workshop, because now I knew we were working together.

Perspective is a powerful force. It changes our attitude, our behavior, and the physiological reactions in our body. When we are willing to stop and consider the other person's perspective, we begin to work together. This is true whether we are conducting a workshop, as I was, or trying to put a child to bed.

WHAT IS HAPPENING "INSIDE" MAKES A DIFFERENCE

When your child doesn't sleep, it can feel as though her behavior is intentional. Why, you may wonder, is she doing this to me? Why is she goading me and disturbing everyone around her? This perspective leaves you feeling angry and helpless, ready to fight with your child or to shut the door and walk away from her. The reality is that when your child isn't sleeping, it isn't about you. Rather it is a reflection of what's going on inside of her body. When she doesn't sleep, it's not because she won't, but because she can't.

Think about your own restless nights, when sleep eludes you. Tossing and turning, you find yourself checking the clock every two hours, your dreams leaving you troubled and tired. You do not choose for this to happen to you. Rather, something is on your mind, your body humming with energy. As a result, you do not sleep because, like your child, you can't.

SLEEP WILL COME WHEN YOUR CHILD'S BODY IS "READY" FOR SLEEP

Awake and asleep are opposite states. The most vulnerable point in your child's ability to sleep is when the brain must "switch" from one to the other. Sometimes that shift appears to be effortless. Your child (or perhaps your neighbor's child) simply lies down and goes to sleep.

Other times, the "switch" can be almost violent, a crashing of systems done only in protest, leaving you feeling lashed by its force. It's the unpredictability of what will happen that can drive you wild. But it doesn't have to be that way. By recognizing what's occurring inside your child's body, you can predict how well the "switch" will work, and, as a result, take the steps necessary to keep it working smoothly. Bed- and naptimes do not need to be dreaded events.

PREPARATION IS THE "KEY"

When your child's body is calm and set for sleep, the "switch" from awake to asleep moves easily. It's when your child's body is agitated that the "switch" gets stuck, and he can't fall asleep. But, once again, a child can rarely tell you with words that his body is too wired to sleep. Instead, he shows you with his behavior and his cries. Watch closely and you'll see it. Take note and you'll feel it. Listen carefully and you'll hear it. No matter how old your child is, when his "switch" to sleep is disrupted, what you will encounter is motion; a restlessness verging on a frenzy.

When I first met five-month-old Jacob, he was a notorious "catnapper" who screamed interminably before collapsing into a fitful nap for thirty minutes, whereupon he would promptly awake, shrieking again. His parents tried feeding, rocking, bouncing, and walking him. Finally, in complete exasperation, they laid him down and stepped away. Nothing worked, and so, they called me. We agreed to meet at naptime.

Jacob watched me carefully from his father's lap, his deep, dark eyes alight. Slowly, I moved toward him, smiling. I reached my hands out as though to take him, but stopped before I touched him. "May I pick you up?" I asked him. His eyes darkened as he thrust his feet out and shoved his body back into his father's chest. I stopped and pulled my hands back. "All right," I replied to his clear message. "I'll wait. Let's just talk." He rewarded me with a smile, smoothly cycling his legs and gurgling in response to my oohs and ahhs.

A few minutes later, his eyes dulled and he began to squirm. His cheeks went slack, as fatigue overpowered his body and he began to fuss. His mother took him then, placing him in a soft carrier strapped to her chest. Facing her, he slipped down and took her breast. Briefly, he quieted and fed, then came off the breast. His head slumped to one side in fatigue as his eyes fought to remain open, flaring wide then drooping as though in a valiant attempt to stay alert. A momentary lull hung in the air as he teetered between wakefulness and sleep. We waited, almost holding our breaths. All of a sudden, the sun broke through the clouds, shooting a streak of bright light into the room.

Jacob jerked his body, tensing. Wails erupted, hard, angry screams. Arms flailing, he wouldn't—couldn't—take the breast when his mother offered it to him again, and instead pushed himself away. Beads of perspiration sprang up on his forehead. Cheeks flushed. A mere shaft of light had unsettled him, disrupting his "switch" to sleep.

Mom stopped walking and bouncing, as she had been doing, took him out of the carrier, and removed his sweater. Knowing that sometimes a change in our voice can help to calm the brain, I started to sing to him, "Hush little baby, don't say a word." He stopped and turned toward me alertly. "It's all right," I told him, "you can relax. You can sleep."

Little by little, a pink glow replaced red splotches on his cheeks, his breaths slowed as his body relaxed and folded into his mother's. His eyes rolled. Rubbing his forehead gently, his mother kissed his eyes shut. We held our breath once more in anticipation of the "switch." It was there, and then, in an instant, it was gone, shattered by a loud guffaw in the next room.

This time, Jacob lost it completely, screaming at the top of his lungs, kicking, thrashing, his body out of control. Mom attempted to sing the lullaby. It didn't work. She turned to me with questioning eyes.

"It's not you," I reassured her. "He's exhausted, and he's a very sensitive little guy. He needs us to help him relax, so that he can shut down and go to sleep."

I asked her to put him back into the carrier, tuck his arms inside, and place a light receiving blanket over him to help block out the light. And then I advised her to slow her movements and sway gently from side to side.

Soon his cries lost their intensity. The breaths between the cries grew longer. Fortunately, at that moment, the fan on the furnace clicked on. Rather than alerting him, the dull, steady hum drowned out everything else with its monotonous drone, and Jacob went to sleep. Thirty minutes later, he awoke again. "The catnap," his mother sighed in frustration. "He does this all the time."

I listened to his cries, verging on that note of despair. "He's not ready to be awake yet. Let's see if we can help him shift back into sleep."

And so, once more, she began to move slowly, quietly swaying, with the light blanket still over him. The cries subsided, and he tumbled into an exhausted sleep. Jacob had been too on edge to sleep and, as a result, his "switch" had been stuck. Despite his body's need for sleep, it took more power to soothe him than he had available. He needed our help. Only when his body was untroubled and his brain quiet, could he "switch" to sleep.

OLDER CHILDREN ARE AFFECTED AS WELL

The ability to fall asleep isn't just an issue for babies. Older children and adults also struggle when their bodies are tightly wound up and humming with energy. Seven-year-old Tommy no longer screams in his parents' arms, but on the "bad days" he, too, labors to fall asleep.

"Before we even start bedtime, I often know it's going to be bad," his dad told me. "On those nights, he has spent the evening running around, tackling people, throwing things, and demanding to play tag, or hide-and-seek. It's as though he's trying to keep himself awake, flitting from one thing to another, talking and singing. When he lies down, he's wiggling and fidgeting. If I pull up his covers, he sits up and throws them off. Two seconds later, he pulls them back over himself. He keeps getting up, selecting another stuffed animal to sleep with, or asking for another drink of water, another hug. He wants one of his action figures, but if I give it to him, it's not the right one and he needs another. He just can't settle, yet he doesn't want me to turn off the light. It's as though he's spouting energy. Minutes after leaving his room, I hear the thump, thump, thump of his footsteps coming down the hallway. I know what's coming next. 'I can't sleep. I feel sick.'

"This can go on for an hour or two. Sometimes, I swear, he's just baiting me. But, finally, I give up and sit with him. Only then does his body begin to relax and he falls asleep."

When your child doesn't fall asleep, even though you know he's tired, it may feel as though he's plotting against you, but he's not.

THE THEORY

Throughout the day, your child's body moves through different states of well-being. How easily the "switch" to sleep works depends upon what state he is in.

In class, I drew a line down the board and, with a green marker, wrote to the left of it: CALM ENERGY, CALM TIRED, and RESTORATIVE SLEEP. On the right, with a red marker, I wrote: TENSE ENERGY, TENSE TIRED, and CRASH SLEEP. It looked like this:

Calm energy	Tense energy
Calm tired	Tense tired
Restorative sleep	Crash sleep

Then I turned to the group.

Imagine for a moment that you are standing on this line. If you step to the right, you move into the "red zone." Robert Thayer from California State University of Long Beach calls this the zone of tense energy and tense tired. Tense arousal is a warning system that turns the body "on," elevating stress hormones and keeping it alert. Thoughts swirl through the brain as the heart races. Literally, the body fights to stay awake.

Step to the left, and you have entered the "green zone." Robert Thayer calls this side calm energy and calm tired. While in this zone, your child's body is calm and relaxed; his heart is beating in steady, deep thuds. All systems are working smoothly and, as a result, your child shifts easily and quickly into sound sleep.

By recognizing the differences between these two zones, you can take the mystery out of whether or not your child will easily fall asleep. Observe closely; your child's actions and your own words can help you identify what zone you are in.

RECOGNIZING THE "GREEN ZONE" OF CALM ENERGY AND CALM TIRED

On the board, I wrote: *What behaviors do you see when you and your child are in the "green zone" of calm energy and calm tired?*

Paul slouched in his chair. As usual, his baseball cap was turned backwards, but on this day, there was a twinkle in his eye. "Is calm energy when a child is full of energy but not wired?"

I nodded.

"He can stick to a project without asking for help every five minutes or needing a drink of water. You hear yourself saying 'good job!' "

I nodded once more.

"Bedtime is a breeze. He's tired, but not cranky. There aren't any big arguments and he falls asleep quickly?"

"Yes!" I replied, pleased that he'd been so observant.

He looked at me thoughtfully, and then quipped: "That's how I imagine other people must live." The others laughed.

Sometimes it has been so long since your child's body or your own has been in a state of calm energy that you've almost forgotten what it feels like. That's why it is so important to begin to notice, because it is in this state that the "switch" works almost effortlessly.

CALM ENERGY LEADS TO CALM TIRED, WHICH RESULTS IN THE "SWITCH" WORKING SMOOTHLY

Calm energy and calm tired are pleasant states. It's as though all systems are aligned. There's liveliness in your step and depth to your patience. You are comfortable in your body and feel good.

Your children feel it, too. On days filled with calm energy, the kids are busy yet focused and productive. While they might become upset or frustrated, they can be talked through the issue and easily comforted. As evening draws near, the energy drops, yet irritability doesn't become an issue. Instead, they move into calm tired. They're content and, most important, ready for sleep. Robert Thayer states, "Calm tiredness produces the deep sleep of the gods."

The reason sleep comes so easily when your child's body is in a state of calm tired is that he is untroubled. Stress hormones are not flowing through his body. His brain is telling him he is safe, he is secure, and it is time for sleep.

Helping your child to stay in the "green zone" is your goal. But sometimes life pushes you over the line into the "red zone." It's essential that you recognize when it happens.

RECOGNIZING THE "RED ZONE" OF TENSE ENERGY AND TENSE TIRED

"What happens when your child moves into the 'red zone'?" I asked my class. Natasha pointed to the door. "I want to run away. I've never actually done it, but there have been times I've laid the baby in the crib and left him there for a few minutes because I couldn't deal with it anymore."

Sarah sighed. "It's so frustrating. Inevitably, it happens on the nights my husband is traveling. All day I've been waiting for those few minutes when I can sit down on the couch and not do anything, or finally get the time to do what I've been waiting to accomplish. But not five minutes into it, I hear those little footsteps on the stairs and then 'MOOOOOM.' My blood pressure shoots right up and I want to YELL!"

You will know that you are moving into the "red zone" when you want to flee the situation or hunker down for a good fight. You can feel tense energy in the back of your neck, the nagging crick that won't go away. Listen carefully, and it will appear in your voice as you shout at the kids, "Settle down!" "Stop it!" or "How many times do I have to tell you, no!" It's in your movements as you rush from one thing to the next, an out-of-sync energy pulling your body in opposing directions, leaving you hot and sweaty. And it scatters your thoughts, leaving you unfocused and frustrated as you tap your toe in impatience and reach for the chocolate.

You'll also see it in your child's behavior. If he's irritable and getting into trouble for not listening, for hitting, or for throwing, you know he's moved into the "red zone." When his motions are frantic and he can't settle down, tense energy and tense tired are building. If he's begging to watch television and craving sweet snacks, it's another sign, and when he starts pulling on his shirt, biting his nails, sucking his thumb, or fidgeting, it's a clear-cut signal that tense energy is taking over.

TENSE ENERGY LEADS TO TENSE TIRED AND "CRASH" SLEEP

When your child's body is in the "red zone," the "switch" between awake and asleep gets locked in the awake position. He remains alert, ready to "fight or flee." It's an inborn survival strategy.

Ultimately, however, the force of exhaustion will overwhelm him, and his body will collapse into "crash" sleep. Researchers at the University of Pittsburgh have discovered, though, that "crash sleep" has a different structure than sound sleep, and doesn't have the same restorative, energizing effect on the body. As a result, a child whose body is in a state of tense arousal may finally crash, but he will awaken frequently during the night, experience fewer episodes of deep sleep, and arise in the morning irritable and still worn out.

Exhaustion creates more tension, pushing your child further into the "red zone." More tension increases the difficulty of falling asleep. Lack of sleep fosters misbehavior. The cycle feeds itself; the more tense and out of sync the body, the greater the resistance to sleep.

YOUR CHILD DOESN'T CHOOSE TO MOVE INTO THE "RED ZONE"

It is important to remember that your child does not choose to move into the "red zone." It's life that can put him there. No one is exempt. That's why it is so important for you to recognize the "red zone" and its impact on your child's ability to fall asleep. When you are aware of it, you can take steps to protect your child from moving into it, and create strategies that minimize it when avoidance is impossible.

In order to do so, you need to be aware of the triggers that send him there. There are many, but in the chapters ahead, I will help you look at three key culprits. They are:

- Tension generated by emotions, especially distress and excitement (even for infants)
- Time—the daily routine or lack of routine can be innocently upsetting your child's "body clock," making it more difficult to sleep
- Temperament—your child's own genetic "wiring" (Some children, because of their inborn temperament, are more easily pushed into the "red zone" and find it more challenging to get out. These children, our most sensitive sleepers, especially need our help.)

All three of these culprits impact what's going on inside your child's body and brain.

While you cannot make your child sleep, you can help him maintain a state of equilibrium. That balance makes it more likely that he will spend his days in a state of calm energy, his evenings in calm tired, and his nights in deep sleep.

EXPOSING THE CULPRITS THAT CAN KEEP YOUR CHILD AWAKE

SIX

TENSION TRIGGERS

Is your child too tense to sleep?

"Oh my, it's true! The busier I get the worse my daughter sleeps and the more she demands to be held."

—Raelene, mother of four

Traffic report," announced the disk jockey as my clock radio sprang to life. "Expect fog in low-lying areas, delays at the 494 and Highway 100 exchange. Allow extra time going home tonight. It's liable to be rough, with a forecast of 3–6 inches of snow."

I lay in bed, musing. Wouldn't it be nice if we could wake up each morning to a "tension" report on our child? I can hear it now. Expect grogginess this morning, low energy levels with difficulties getting out the door and onto the school bus. Be prepared tonight. There's a substitute teacher today, notorious for her control tactics and screaming. Anticipate high tension levels in late-afternoon hours.

Alas, we're left to our own devices to figure out the unexpected changes, upsetting events, and frustrations of the day that can raise tension levels. But we don't have to be blindsided by it. Just like the traffic reporter checking her morning monitors, we, too, can prepare for the day, anticipate some of the potential problems, and plan accordingly.

TENSION—THE INVISIBLE CULPRIT

When tension levels are high, your child struggles to "regulate" his body. Unfortunately, he doesn't arrive on this earth with an understanding of tension or its impact. He doesn't know why he is wild with energy when you desperately need him to be sleeping. He doesn't know why he is screaming when he's so exhausted. Nor does he recognize that he's begging for one more book at nine o'clock at night because the home run he hit at six-thirty still has him too excited to sleep. It just happens.

And that is the challenge of managing tension. Many of the "culprits" that create tension are part of our everyday lives. They're so "normal" that it's easy to miss them, but when we fail to recognize them, they can creep up on us, leaving everyone in the family wide awake and wondering why. It is our job to help our children identify tension as it occurs, give them words to use, and teach them strategies for managing it throughout the day so that at night they can sleep. Ultimately, they'll take over the task themselves. The task requires a bit of sleuthing, learning to notice the things that generate tense energy, and a tad of resolve on your part, to take preventive measures.

Learning to identify the everyday "culprits" that create tension

When my friend drives to work, there's a man who often stands on a walkway overpass, holding a protest sign. Inevitably, traffic slows as drivers glance up to read it, which ultimately causes a tie-up. As a result, my friend has taught herself to begin looking miles before the overpass for brake lights far ahead or slightly slowing traffic that indicates to her that once again the man is on the overpass. When she sees these signs, she takes a different route and avoids a significant delay. But if she is talking on her cell phone or thinking about other things, she often misses the signs until it's too late. She can no longer exit and gets caught in the jam. The same is true as we work with our kids.

Recognizing how distress and excitement trigger tension

All too frequently, your child's reluctance to go to bed is actually a reflection of how her world is feeling at that moment. Sleep is very sensitive to our emotions. And while it is well documented that emotions can disrupt sleep for adults, what is not as well known is that they can also disturb the sleep of children—even infants. That's because emotions "arouse" the brain and body. As a result, our muscles tense, preparing us to take action.

While you can't predict *every* possible emotion that leads to tension, what I've discovered from working with families and by reviewing the research is that there are two key categories: emotions that create distress and those that generate excitement. Within each of these categories, I've identified five common triggers you may be dealing with every day that can be leaving your child "on alert," too distressed or excited to sleep. Once you know what to look for, you won't be surprised, or feel out of control. They are:

Common triggers for Distress	**Common triggers for Excitement**
Parental stress	Overstimulation
Separations	Overscheduling
Upsetting events	Anticipation
Major life changes	Growth spurts
Lack of sleep	Competition and pressure to perform

When you learn to anticipate the "culprits," you can take steps to help your child relax. As a result, you'll discover that the intensity of the emotions won't overwhelm him. The meltdown tantrums will disappear, and he'll be ready at bedtime for sleep. Discovering them is like solving a challenging puzzle. You will be delighted with your success.

The previous week, Sarah had lamented that her son Jason often had difficulty falling asleep at night. "I feel so frustrated and helpless. Isn't there something I can do?"

"Be aware of things that distress or excite him," I had offered. "When you see him becoming tense, stop, connect, and calm him. By catching the emotions when they're 'little' and easier to manage, you will prevent him from going into the 'red zone' of tense energy."

The next week Sarah bounded into the classroom. "It worked! Jason had a reading assignment. The questions required him to analyze the material, like who was the loneliest character. He was completely melting down trying to complete the assignment. I tried asking him questions, but that didn't work. I started to get angry, and then it dawned on me. He's really distressed, scared that he won't be able to do this assignment. I had to think what might calm him, and decided he needed a break, so I told him to go ahead and get on his baseball uniform for practice. Then I encouraged him to bring his book and read to me while we drove to the field. After a few paragraphs, I stopped him and asked, 'What do you think that little boy was feeling?' He paused, and in the rearview mirror I saw his eyes light up.

" 'Wow, he's really missing his mom. He's the loneliest!'

"I could actually see the tension melting from his body. And when we got home, he was able to quickly complete the assignment. He would have lain awake worrying for hours if he hadn't finished it. Instead, he went right to sleep."

When you recognize your child's distress and understand its source, it's easier to stop yourself from hurling out the threats that will only catapult him into a steeper downward spiral. Instead of pushing harder and escalating his reaction, you will be able to diffuse it. Responding to your child's distress during the day keeps him sleeping at night. So, let's tune in and get a picture of your child's tension level.

Getting a picture of your child's tension level

Following the prototype I use with the families I meet with one on one, I'm going to ask you to do a "life check." As you review each tension trigger, think about what's been happening in your child's life during the last six months. What triggers are sending your child over the line into tense energy? Every child will respond differently to stressful experiences, but if you watch carefully, your child will tell you with her behavior when she is feeling overwhelmed and needs your help calming down.

IDENTIFYING THE TRIGGERS THAT DISTRESS YOUR CHILD

Sometimes it's difficult to imagine what could distress a young child, especially an infant. But anything that disrupts a child's sense of safety and security is likely to lead to higher tension.

Your stress can create tension

Anthropologist Mark Flinn from the University of Missouri has been studying children living on a remote tropical island for more than thirteen years. He conducts his research by asking kids to spit into a cup. While this activity may appear a bit odd and unpleasant, saliva contains a hormone called cortisol, which is produced in response to stress. By testing the saliva, he can identify what's going on inside each child's body. What he discovered is that children's (even infants') stress levels peak when the key adults in their lives are stressed. What may seem inconsequential to adults—a fight between Mom and Dad, Grandma fretting about bills, or Mom leaving on a business trip—causes a child's cortisol levels to rise.

You might be thinking, "My two-year-old doesn't understand the flip of my stomach when I open the Visa bill." That's true. She doesn't, but she senses it. It appears that, without meaning to, you can communicate your stress to your child via your touch, voice tone, and gesture. When you slam the door, throw down the car keys, or yell, the force and tone convey to your child that something is amiss and that he needs to be on alert. Immediately, stress hormones are released into her body. Your stress also preoccupies you, making it less likely that you'll pick up your child's cues and respond patiently. The result is a child who feels more anxious and insecure and, as a result, fights to stay awake.

This, of course, only adds to your frustration. When you need your child to sleep the most, she doesn't. While this may appear to be very bad news, it's actually not. If you are battling over bedtime and the misbehavior which results from the residue of sleep deprivation, ask yourself, what is my stress level? Are you worried about finances, or about losing a job? Has conflict at home or work escalated? Are job responsibilities expanding? Are you feeling trampled by the demands of young children? It's the little things that can activate the body's stress system, leaving the whole family squabbling in the middle of the night. Once you recognize the relationship between your stress and your child's misbehavior and name it, your heart rate will slow, you'll be calmer, and your child will respond accordingly. You will have just turned the corner toward better sleep.

Separations can create tension

Whether it's merely an evening with a sitter, or a long-term separation due to a divorce or a parent's military deployment, separations can create tension for your child. This doesn't mean that you should never leave your children. It's simply a "heads-up," so that you can consciously think about the things that will help them feel safe and secure during your absence and get the sleep they need. Separations are such a normal part of our lives that without this awareness, the tension that they generate can catch us off guard.

Business travel is a common experience for many employed parents, including me. Yet studies show that every time a parent leaves—even though it's a routine event—a child's cortisol level rises, leaving him cranky and more difficult to get to sleep for at least the first day or two before things start to settle down again.

Megan Gunnar's team at the University of Minnesota has also discovered that toddlers in child care demonstrate progressively rising stress hormone levels throughout the day. The larger the group, and the longer the day, the more stressful it appears to be for the children. The researchers believe that it is the complexity of the social interactions that make it challenging for a toddler to handle long days in group care. (Preschoolers and older children who are more adept in social situations do not demonstrate rising stress levels.) This doesn't mean that child care is bad for your toddler's well-being, but simply that if your child is struggling to sleep, there may be a link.

Even separations from siblings and peers can disrupt your child's sleep patterns. Studies demonstrate that infants and toddlers who "graduate" from one room to another at child care show increased activity, longer time to fall asleep, more crying prior to falling asleep, and decreased amount of sleep during naptime for the week preceding and following the change.

Now, this can feel like a double bind. How are you supposed to manage your own stress or hold down a job if separation creates tension for your child? Once you put your finger on the "culprit," you can explore ways to manage it, like shortening your child's day at child care, when you can, or planning extra calming strategies for the evening, when you can't.

When I mentioned in my conversation with Jamie that separation can be a "trigger" for tension, her eyes lit up. "My husband travels every week and always has. I've never stopped to think about it. The kids are with me, what more could they need? But it's a fact, on Mondays, the kids are always fussier. They get on each other's nerves and fight much more, and I can't get them to bed. By the end of the week, things have usually settled down again, and I forget about it. I'm going to pay attention to this."

The next week, she nearly jumped out of her chair in excitement. "Last week, I decided to see if I could make Monday better. I explained to the boys that it was harder for them to sleep right after Daddy left. We made a plan. At bedtime, I'd read to them, and then I'd give them a fifteen-minute massage. Normally, on Monday nights, they're up until ten o'clock, but this week, they were down by eight-fifteen."

So, when the frequency of bedtime battles and power struggles increases, note if there has been a separation. Have the hours your child spends at day care increased? Is your child switching back and forth between Mom's and Dad's homes after a divorce? Did your child spend more hours with a new caregiver recently? Did a good friend move away? And don't forget to consider "graduation" from one class to another, or a teacher going out on maternity leave. It's much easier to soothe a child when you understand that tension created by a separation is making your child cranky and uncooperative.

Upsetting events can trigger tension

It doesn't have to be a traumatic event such as an accident or major illness to increase arousal and agitation for your child. Getting lost in a store, being held down for a painful medical procedure, experiencing a bad storm, hearing a terrifying news story, or having a teacher or coach who yells and shames, can be enough to keep your child awake at night for days, even weeks. That's what happened with Tola.

Ten-year-old Tola had always been one of those kids who fell asleep easily on her own, but suddenly she was insistent that her parents stay with her until she did. Puzzled and frustrated, they called seeking help. During my visit to their home, I noticed that the television was left on in the background. The thought occurred, If it bleeds, it leads in the news. I turned to Tola and asked, "By any chance, have you heard any news stories that frightened you?" "Oh, yes!" she exclaimed. Eyes wide with fear, she proceeded to describe the kidnapping and brutal murder of a young girl. The event had been the lead story for days.

It can be difficult to know what will significantly upset a child. During the last six months, has your child or family experienced any painful or distressful event? The residue may be lingering in your child's body, pushing her across the line into tense energy.

Major life changes can create tension

Major changes can also pose a problem. A move, a new baby, a divorce are obvious creators of tension, but what may not be as obvious are the little changes that actually have a big impact on tension, like living through a home-remodeling project—especially a kitchen—which disrupts your entire family's sense of order and predictability and drains your pocketbook. Switching beds or bedrooms, going on a family vacation, the start or end of a school year, or even the shift to daylight savings time can impact your child.

The demand for my private consultations surges in September. I can predict that three weeks into the school year, I'll be getting calls for help. That's when the honeymoon period ends, sleep deprivation kicks in, and behavior problems rear their heads. Springing forward for daylight savings time is busy, too, and so is the end of the school year. It's not that children check the calendar and say to themselves, "Ah ha! Now we'll drive our parents wild." Instead, it's the challenge of shifting from one routine to another that can keep them overly alert. Change, whether distressing or exciting, can leave your child too tense to sleep without some extra help.

Lack of sleep is a stressor in itself

Ironically, the less sleep your child has the more stress hormones his body releases to keep him going. If your child isn't sleeping or behaving well, think back on the events of the week. Did you have to wake him up from a nap? Did he skip a nap or stay up late for a special event? Did he spend a restless night in a hotel or at a slumber party? If these things occurred, you can assume that your child is experiencing high tension.

IDENTIFYING THE TRIGGERS THAT EXCITE YOUR CHILD

Positive emotions like excitement and anticipation can also create tense energy. One of the biggest challenges today is avoiding overstimulation.

Overstimulation can create tension

Lights, noise, crowds, and colors are all sensations that can stimulate the brain. Some children seem to easily block those sensations and drop off to sleep in the midst of them. Others get revved up and just can't fall asleep. But high levels of stimulation are the norm for most families, and, as a result, it is easy to miss this as a cause.

What does overstimulation look like? Deanne gave a good example in this anecdote about her six-year-old daughter Kia.

"Last night, when I came home, my husband, Bob, was in the kitchen, which opens into our family room. He was watching the news on a small television that hangs on the wall above the cabinets. Kia was lying on the couch, listening to a CD on a headset. It was playing so loudly I could hear it while standing fifteen feet away. She sang along to the words, seemingly oblivious to the commotion around her. Becca, our two-year-old, was dancing in front of the big-screen television AND listening to her own CD on the entertainment center. Every two seconds, she was running to Bob, shouting, 'Turn it up, Daddy, turn it up.' I couldn't face it. I turned around and walked back out the door!"

Do a life check. Did battery-operated toys arrive as gifts for your newborn? Are surround sound and big-screen television frequently turned on in your home, especially in the evenings? Does your child watch a video while driving with you through rush-hour traffic? Have you ever noticed that, after a day of shopping, your child can't sleep?

Stop, look, and listen. How many different sensations is your child's brain trying to process at once? Does the stimulation level in your child's life leave him cringing, too tight to sleep?

If your child is especially sensitive to stimulation, it doesn't mean that you should never go to an amusement park for fun, or a restaurant for dinner. It's just a reminder that if his day has been filled with hours of television-watching, crowds of people, and a barrage of stimulation, it's likely that he'll need more help settling down for the night. He's not ungrateful for the good time. He's wired.

Overscheduled days can be keeping your child awake

Sometimes it's the pace and sense of rushing that can be keeping your family awake. Even when you've been looking forward to the activities and thoroughly enjoy them, there's a line where you and your child cross from calm into tense energy.

After talking with a family about the importance of slowing down enough to sleep, I received this e-mail: "If you recall, you were with us on a Friday afternoon. Our plans for that weekend were packed with activities, including a basketball game, a visit to Santaland downtown, dinner out, and a play at the children's theater. Sunday was scheduled to be just as hectic, with a birthday party, a church musical program, and dinner with friends. After your visit, my husband and I sat down and looked at that schedule. We immediately realized that it was too much, and that it also explained why Nathan was melting down so frequently and unable to sleep. The fog lifted. We could fix this!

"We decided to visit Santa locally rather than going downtown. That gave us extra time, which allowed us to spend the afternoon at home, doing quiet activities and eating supper before heading to the theater—great day and night! Sunday was trickier, because the events were back to back. But I told Nathan I'd stay with him at the party—which meant it was less stressful for him—and when his brother's part of the musical was finished, I took him home for a thirty-minute break. Once again, he made it through and fell asleep that night feeling very proud of himself."

If you can name it, you can change it. It's hard to imagine that plans that are supposed to be exciting may actually be revving your child up. But when your child constantly feels rushed, he moves into tense energy to keep going. Is the pace of life leaving both you and your child with the sensation that your shoulders are pulled up to your ears? Are you awakening at five-thirty a.m. your brain eager to continue the race, fueled by the cortisol coursing through your system? Are you short on sleep? Often we become so accustomed to this level of tension that we are not even aware of it.

Take special care to pay attention to the needs of a younger child who gets toted along. The toddler who, in the middle of the shopping center, starts to shout, "Booby, booby," and tries to lift your blouse so that he can nurse is telling you that he's had enough. And the preschooler who dashes to the door, trying to run out, is letting you know it's time to go.

The stress of a too-busy life can get you and your child not only during the day, but at night as well. Recognizing this allows you to find the balance between a busy, yet satisfying day and one that leaves everyone in a frenzy.

Anticipation can keep your child awake

Sometimes it's a fun project or an event that you and your child are looking forward to that is keeping everyone awake. For Noah, it was a birthday party.

"Noah had a birthday party right after school," his mother explained. "He was excited and just a little uneasy about going there after school. The night before the big event, he pulled the twenty-one curtain-call trick. He needed a drink of water. His feet hurt. Would I rub them? His pillow wasn't right. I was getting really frustrated with him, knowing that the next day was going to be extra busy, but then I realized he was so excited. Suddenly I had more patience with him and was able to help him settle down."

Birthday parties, vacations, holidays, and family gatherings are supposed to be fun, but sometimes they're hard on our sleep. According to a recent survey by the Gallup Organization, 76 percent of American adults report losing sleep between Thanksgiving and New Year's Day, 49 percent of them losing three or more hours of sleep a week. Almost half of those polled, 48 percent, reported experiencing some degree of stress-related sleep disturbance during the holidays. It's not too great a leap to assume that children are affected in the same way.

Anticipation can also be more about anxiety. If your child is worried about whom she will sit with at lunch, or who will play with her at recess, she may lie awake, fretting. As a result, you can't get her out of bed in the morning.

Whether it's an exciting project or a worrisome event, the anticipation of it can leave your child edgy and unable to sleep. What's essential to recognize is that she needs your help calming down, not your anger.

GROWTH SPURTS CAN LEAVE YOUR CHILD TOO EXCITED TO SLEEP

Who will sleep better, a six-month-old baby or a ten-month-old baby? The answer is the six-month-old. Psychologist Tom Anders found in his studies that children nine to fourteen months old wake more frequently than six-month-old infants. The reason, he believes, is the huge surge in physical development at this stage. It's during this period of nine to fourteen months that most tiny toddlers begin to pull themselves up to standing, and begin walking. The joy of these new skills raises arousal levels and so enthralls the child that even in the middle of the night he wants to practice.

It's not only infants who experience growth spurts. The older toddler who suddenly has a burst in his vocabulary, the four-year-old who discovers an imaginary friend, and the nine-year-old who wakes up one morning suddenly realizing that the events she's heard about on the news are actually happening in her community, are also experiencing developmental milestones. All individuals, including adults, experience significant changes in emotional, physical, and cognitive growth. The difference is that for young children, growth spurts occur about every six months. Through adulthood, we slow down to one in every seven years.

So, if your child is waking in the night or battling to stay up, ask yourself, is she within six weeks of her birthday or half birthday when growth spurts tend to occur? Or have you noticed any significant change in her skills?

That's what five-year-old Isaac's parents discovered. Isaac wasn't sleeping. He, too, had been a good sleeper, and, once again, the family was stumped until they started comparing notes with the other caregivers in his life. Then they realized that Isaac suddenly knew his letters and numbers and was trying to read and write. His drawings now included great detail instead of his old scribbles. In fact, all of his skills were taking off at the same time. This learning binge was creating tension that kept Isaac awake.

What skills is your child working on right now? What is he able to do that he couldn't do six months ago? The quest to grow may be keeping him aroused.

And sometimes it's not even his own growth spurt that is causing a problem for your child. When a younger sibling starts to crawl, walk, or talk, he "invades" the space of the older child. The "invasion" can be upsetting enough to keep him riled up and awake.

Competition and pressure to perform can leave your child lying awake at night

Competition and the pressure to perform is the American way, but it can wreak havoc on sleep. The pressure to complete homework and compete at school may be obvious to you, but a competitive game before bed may also create a sleepless night.

Ten-year-old Ike and eight-year-old Ben liked to play basketball. Often, Lisa, their mom, a former Big Ten collegiate player, would play with them. After dinner, it wasn't uncommon for the trio to head for the driveway and a full workout before bedtime. It was only after attending class to figure out how to reduce the sibling struggles that Mom realized that Ben, who often "lost" the game, had great difficulty falling asleep afterwards. Short on sleep, he was unbearable the next morning.

Competition and pressure stir up the brain. That doesn't mean that your children should stop their competitive pursuits. The key is to consider the timing of them, and if you can't control the timing, to plan for wind-down afterwards. Today, Lisa and the boys are still playing basketball, but now they play *before* dinner, so that there's time to unwind. If they play after dinner, they end with relaxing dribbling drills rather than the "winning" basket.

Now go back through each of the "culprits" and identify those that are significant sources of tension for your child. Pay attention to them. Notice how the tension they generate is linked to your child's misbehavior and inability to sleep well.

TRIGGERS FOR DISTRESS

	Not significant	Significant factor
Parental distress		
How distressed are you?	______	______
Are you experiencing financial pressures or increased responsibilities?	______	______
Have you experienced more conflict with key people in your life?	______	______
Separation		
How much separation has your child experienced lately?	______	______
Has your business travel increased?	______	______
Has a sibling gone off to school?	______	______
Has a best friend moved away?	______	______
Has the length of time in child care or with sitters increased?	______	______
Upsetting events		
Has your child been exposed to	______	______
potentially upsetting events?	______	______
Has there been a major storm, or natural disaster?	______	______
Did someone in your family experience illness?	______	______
Major changes		
How many significant changes has your child experienced in the last six months?	______	______
Has there been a divorce or a move?	______	______
Has your family recently traveled?	______	______
Losing sleep		
Was your child unable to nap?	______	______
Did bedtime end up later because of other commitments?	______	______

TRIGGERS FOR EXCITEMENT AND ANTICIPATION

	Not significant	Significant factor
Overstimulated		
Has your child been exposed to a high level of stimulation?	______	______
Does your child easily become overwhelmed in crowds or busy public places?	______	______
Is the television on as background noise?	______	______
Overscheduled		
How busy is your child's day?	______	______
Is there a constant sense of "rushing"?	______	______
Anticipation		
Is anticipation keeping your child awake?	______	______
Is your child looking forward to a holiday or other special event?	______	______
Does your child "worry"?	______	______
Growth spurts		
Is your child demonstrating new	______	______
skills, or is within six weeks of birthday or half birthday?	______	______
Competition and pressure to perform		
Is your child experiencing high pressure or lots of competition?	______	______
Does your child have a project due?	______	______
Is your child participating in competitive activities in the evenings?	______	______

Sometimes it's the cumulative effect that matters

Often, children seem to manage well the emotions created by a move, a new baby, or a teacher gone on maternity leave—until six months later. Then, suddenly, their coping skills deteriorate, and the child who has been doing well is not any longer, and, as his behavior worsens, so does his sleep. It's often difficult to imagine that an event that happened six months ago can disrupt your child's sleep and lead to major behavioral problems. The reality is that it can.

When you are able to tune into the "culprits" that are creating tense energy in your child's life, you won't feel so out of control. As a result, you'll respond more empathetically, recognizing that your child is not trying to be difficult. Your awareness will also allow you to be kinder to yourself. You are not a bad or ineffective parent. It's tension that is keeping your child on alert, unable to sleep and acting up. The "force" is no longer invisible. It's concrete and manageable, and you are now ready to take the steps to reduce it, so that everyone can sleep.

SEVEN

EASING THE TENSION

Helping sleep come quickly

"I always thought of sleep as something to do in our spare time."

—David, father of four

They called her Tara the Terror. It was not uncommon for Tara, the most petite but competitive member of the family, to leap over the stair railing to land on her brother below. The height of the stairs gave her an advantage her size denied her. Just four, she earned her nickname on a daily basis. The challenging behaviors were not limited to daytime. At night, Tara lay on her bed, listening to stories, only to pop back up again as soon as the reading was finished. It wasn't unusual for her to lie awake for two hours before she finally fell asleep. When I met with her family, it was quickly apparent that Tara was chronically short on sleep and running on tense energy. In order to stop the misbehavior and to get Tara some sleep, we had to calm her body, but to do it, we needed Tara to work with us.

Throughout the day, it is normal for the level of tension to fluctuate. What I've discovered in my work and found supported by the research, is that the most resilient individuals take steps to "buffer" and manage tension as it occurs. They don't wait until bedtime to begin relaxing.

If we want our children to fall asleep easily and quickly at night, how we spend the day matters. But in order to "catch" the tension as it occurs, we need a way to talk about it. That can be a bit challenging when your child is little or a preadolescent who doesn't wish to discuss her feelings. That's why I developed the "volcano."

The "volcano": finding a way to talk about tension

I arrived at Tara's home with my black leather tote. In it, I had brought a tray, a wine glass—simply because it's the right size (although I often get teased about always having it handy)—and the ingredients for a "volcano": vinegar (dyed red with a few drops of food coloring for the "lava" effect), baking soda, and four cotton balls. I invited Tara to make a volcano with me. She couldn't resist, and leaned in closely on the kitchen table to ensure a good view.

I explained that I was going to ask her some questions. She could answer each one, or she could "pass." Her brown eyes opened widely as she nodded soberly.

"Can you tell me something that makes you happy?"

A smile lit up her face. "Chocolate cake!"

"Chocolate cake makes me happy, too!" I exclaimed as I handed her the squeeze bottle containing the vinegar and helped her to pour a few tablespoons of it into the glass.

She sat back, waiting for the next question.

"Can you tell me something that frustrates you?"

Eyes rolling upwards, she took a moment to think. "My brother, when he takes my toys!" With my nod of approval, she grabbed the vinegar bottle and added more to the glass.

We continued, going through irritated, excited, angry, scared, and, finally, jealous. By the time we got to jealous, the glass was nearly full, so I handed her a spoon and allowed her to scoop a teaspoon of baking soda into the glass. Immediately, it erupted, a bright red froth of foam bubbling over the top of the glass onto the tray.

"You did it! You made a volcano!"

Gleefully, she clapped her hands.

I leaned toward her. "Do you ever feel like this inside?"

The smile disappeared, as she paused to stare intently at the bubbles. "I feel like that all the time."

I heard a sigh escape from her mom. Under my breath, I whispered, "It's OK. She knows her feelings—that's good."

I handed Tara an oversized cotton ball and asked, "When you're feeling all bubbly inside, like the volcano, what makes you feel all soft and good inside, like this cotton ball?"

She pointed at her mom. "When she gives me a hug."

Mom smiled weakly in relief as we dropped the cotton ball into the glass. The final three cotton balls went into the glass as Tara told me that she liked her dad to read to her, she loved her blanket and her big brown teddy bear. Now the glass was full of vinegar and cotton balls. I handed the spoon back to Tara and let her drop another teaspoon of soda into the mixture. It fizzed, but it didn't boil over.

"Why do you think it didn't bubble over?" I asked.

She pondered for a moment, and then shrugged.

"It's the cotton balls," I explained. "And when you feel as though there's a volcano inside of you, you can ask for your cotton balls. You can say, 'Mom, I'm too bubbly inside to sleep, please hold me.' 'Or, Dad, I feel all fizzy inside, please read to me.' " Her eyes grew big as she soaked it all up. Now she was ready to work with us.

Once your child has a picture of the "volcano" in his mind, he has a tool, a way of expressing what tense energy feels like. So, if your child is three or older, I'm going to invite you to make a volcano with him. Even older children can benefit from this demonstration. (Obviously, you would do it differently, but the idea of "bubbling over" remains the same.) If your child shrugs and "passes," unable to answer the questions, don't fret about it. Go ahead and let him pour the vinegar into the glass. It simply means that you'll want to go back later and work with him, helping him to recognize his emotions, name them, and learn what calms him.

The night after my visit with Tara, I checked my telephone messages. There was one; a single sentence, the voice filled with awe. "She's asleep." It was Tara's mom. Later, when we talked, she explained how at bedtime Tara had started to wind up. "I had that image of the volcano in my mind, so, instead of getting upset with her, I simply asked her, 'Tara, are you feeling the bubbles inside of you?' She stopped, looked at me, and nodded. I stooped down and gave her a hug. I was calm and held her, stroking her back until I felt her relax in my arms. I offered her a back rub. Fifteen minutes later and ninety minutes earlier than usual, she was asleep and I was reading a magazine!"

**In order to manage the tension,
your child needs to hear phrases like:**

"I think your body is wired."

"It looks like your body is full of bubbles."

"Your body is restless."

"It feels like you are humming with energy."

"I think your volcano is smoking."

"I suspect your brain is spinning."

These phrases are often less "threatening" to children, allowing them to open up and talk with you. As a result, when children hear these words over and over again, they can begin to turn them into "I" messages, like:

"I'm feeling wired."

"My bubbles are up!"

"My body is humming."

"The volcano inside of me is smoking."

"My brain is spinning."

By giving them words that are comfortable for them to use, you give them the tools to recognize what's happening inside of their body and to get their needs met appropriately.

After I met with his family, seven-year-old Nathan found these words to be most helpful. It was a big night for Nathan. He'd had three hits and a spectacular save at shortstop that won the game. Coming off the field, he was "high," jumping up and down, grinning from ear to ear. Dashing to the car, he exclaimed to his mom, "My bubbles are really up tonight, Mom!" She laughed with him, and said, "You're so excited. I wonder if it will be difficult to fall asleep tonight." Together, as they drove home, they brainstormed what might relax him, and came up with fifteen minutes of reading after snack and a little bit of massage. Instead of lying wide awake for two hours as he had done in the past, Nathan was sound asleep just twenty minutes later than normal.

Teaching your child what tense energy feels like and giving it a name helps to calm his body. The words help you and your child to recognize what you're dealing with, and, as a result, you stop reacting and instead start developing a plan for success. Managing tension is a team effort. Initially, you, as a parent, catch it and help your child to relax. Ultimately, your child will manage it on his own.

The image of "bubbling over" is one that you can even talk about with infants and toddlers. As you comfort the sobbing infant, you can say, "I can feel the energy inside of you." Or, "You are bubbling." It will remind you that he is having difficulty sleeping, but, sooner than you might ever expect, your toddler will one day surprise you with, "Bubbles up. Hug." You will know that your child is on the road to learning how to manage tension.

Once you have the words to use, you and your child are ready to work together.

Choosing to connect and calm eases the tension

Study after study has found that social ties are critical for managing tension. When we choose to "connect," hormones are released in the system, which counter stress and produce a calming effect. But in today's busy world, finding time to connect can be very challenging.

William Doherty, Ph.D. from the University of Minnesota, states that "The natural drift of family life in contemporary America is toward slowly diminishing connection. Only an 'intentional family' has a fighting chance to maintain and increase its sense of connection over the years." That's why it is so important to deliberately begin the day by thinking of a way to connect with and calm your family. Fortunately, it can be as simple as the look on your face.

EASING TENSION IN THE MORNING

You can set the tone for the day

Waking in the morning is the reverse "switch" from asleep to awake. It's a major physiological transition, and if you have a child who finds shifting from one state to another difficult, waking can be the first step into the zone of tense energy. You can ease that tension simply by the look on your face.

Emotions are contagious. In *How to Negotiate with Kids,* Scott Brown tells the story of a woman whose child always cried when she awoke in the morning. After learning that emotions were contagious, she said, "I realized that every morning I went into her room frowning or scowling. So, I put a mirror on her door. Now, before I go in, I practice a few cheerful lines, like 'Hi, Anna. What a good day we're going to have today.' And I make sure that I'm smiling."

Imagine if your alarm clock was a scowling, shouting monster with bad breath. I suspect you'd quickly replace it rather than endure the surge of tense energy that would rush through your body the moment it went off. Your children don't want to replace you (even if they sometimes threaten to), but they do appreciate it if you brush your teeth before the morning kiss and offer smiles and snuggles instead of shouts and shoves.

You get to set the tone for your family in the morning. So, begin your day with one goal: calm energy. Be kind to yourself, instead of thinking about all of the things you have to accomplish and the things you have no control over, stop and pay attention to how you are feeling. What's your tension level? If you realize there's a kink in your neck, take a deep breath and focus on your breathing. Try to draw your breaths from deep in your belly. As you walk down the hallway, consciously think about placing your foot flat on the floor, stepping softly. Smile; relax the muscles in your face and neck. Lower your shoulders. Move smoothly. Loosen your grip.

Your child will feel the difference and be calmed by your presence and the sense that you are more "attuned." As a result, he will be more cooperative. Minutes won't be wasted in arguments. Even the disappointment of finding his favorite shirt in the laundry hamper will be unlikely to throw him into a meltdown. You'll have more time and feel comfortable trusting that, indeed, you will accomplish the most important things for that day. Checking your tension level is like dropping a "cotton ball" into the system. By doing so, you start the day with a "cushion," helping yourself and your child to manage the surprises and the demands of the day. It really can be as easy as a deep breath, a soft step, or a smile.

Breakfast builds a "buffer"

I remember waking, as a child, to the smell of French toast frying and real hot chocolate warming on the stove. It was always my favorite meal of the day, but somehow, as my children were born and my husband and I were both getting ready to go to work, French toast gave way to a quick bowl of cold cereal and a glass of juice on the run. That is, until the day I visited my friend Kim's home and watched as her family sat down together for a healthy breakfast. Despite the fact that five of them were trying to get out of the house at the same time, they stopped, sat down, and ate together. As I watched them and listened to their conversation, I realized breakfast can create a "buffer." I went home and changed our family's morning routine.

If "We're running late" is your morning mantra, your family is starting the day in tense energy, and your children are going out to face the world with their reserves on "empty." It's tempting to let the kids sleep in until the last moment, but then they're rushing to catch the bus. And as they race through the kitchen, it's likely that you're yelling at them to eat something. So, they grab a smoothie from the refrigerator and gulp it down. They're in the "red zone" of tense energy before they even get out the door.

By sitting down together for breakfast and turning off the television or radio so the world doesn't intrude upon your conversation, your family begins the day peacefully. This provides you with an opportunity to "check in" and monitor your child's tension level and talk about the things that may be disturbing him. Those few minutes are like a shock absorber, supporting you and your child as you cope with whatever life throws at you.

"I have to admit," Leah remarked, "when I heard you talk about sitting down for breakfast, I thought you were living in a fantasy world. But despite the kids' protest that they needed to sleep in as late as possible, we tried it. Now we start the day fifteen minutes earlier. They come downstairs one by one and sit at the table and chat with me while I finish preparations. When everyone is there, I sit down, too. We hold hands and pray, asking God to help them remember what they studied, or whatever else is on their mind. We talk and we eat. They're smiling and chatting. I can't believe how much my daughter, who has always hated breakfast, eats. Now, when they walk out of the door, I can see the difference in their bodies. They're serene."

Breakfast is an easy meal to prepare. If it's helpful, make a menu for the week, including each family member's favorite at least one day. When you create a list, everyone knows what to expect, and you don't have to stop and make a last-minute decision. You can even keep the menu the same every week, if you like.

It's true that you'll have to go to bed fifteen minutes earlier, to gain the morning time, but I suspect that you'll find everyone is more willing to go to bed knowing that they will wake up to a pleasant morning start. Sitting down for breakfast starts your day with a clear message: in this family, our goal is to feel close and calm. Those few minutes early in the morning will actually help your child fall asleep more quickly at night.

MANAGING TENSION THROUGHOUT THE DAY

Planning special moments

Starting your day by spending time together is a very powerful tool to buffer your family from stress. Intentional rituals of connection do not need to end with breakfast. It's important to continue monitoring and easing tension all through the day, so that it doesn't overwhelm your child.

One day, I watched out of my office window as the falling snow played dress-up with the trees, adding skirts to some and puffy sleeves to others. A face was plastered on a tall oak, a sharp nose protruded above white whiskers. It made me think of paper dolls, and loving memories of "Mom Yeadon" flooded me. After school, my sisters and I often visited Mom Yeadon, our next-door neighbor. On her linoleum floor warmed by the oil stove, Mom Yeadon would magically cut paper dolls from old newspapers and help us dress them with clothes cut from the Sears catalogue. No sense of urgency existed in her domain, just lazy afternoons, eating brown-sugar sandwiches and cutting paper dolls, soft clicks of the oil stove our mantra. The mere thought of her relaxed me.

Days later, I was still thinking about Mom Yeadon. I called her and asked, "Why was it so peaceful at your house?"

"Oh," she chuckled, "that was the best part of my day. You know, there's a little bit of a child in each of us, and I think I just enjoyed playing with you." Decades later, those afternoons still rewarded both of us with grins—and a sense of tranquility.

Thinking about my experiences with Mom Yeadon, I wrote a question on the board for my class: WHO WAS THE PERSON IN YOUR LIFE WHO MADE YOU FEEL CALM INSIDE?

The room grew quiet as they dredged up childhood memories. It was Lynn who began softly speaking. "My mother," she offered. "Every morning, until I started first grade, she would hold me on her lap, rocking and reading with me. I missed those mornings after I started school." Others nodded. Robert remembered riding his tricycle in circles while his dad strummed his guitar and sang tunes he created just for Robert. Sarah recalled the sensation of her grandmother's hands, rough from working in the garden, stroking her back as they played games together. Paul simply "hung out" with his father after work, sitting on the back porch, drinking lemonade; the silence comforting rather than heavy. Natasha remembered the lunches her father made her each morning before he left for work, the brown bag decorated with an original hand-drawn cartoon that all the other kids envied. Christine's eyes twinkled, and she sighed contentedly as she described cooking with her parents, listening to them chat about their day, their voices low and melodic as they chopped vegetables, scrubbed potatoes in the sink, or stirred pots on the stove. Simple, everyday events that became special because of the focus, warmth, and love they communicated. Moments of peace, islands of comfort scattered protectively throughout the day of a busy life. Each time you stop to connect, you have the opportunity to "check" the tension level. Your attention calms your child and helps him to manage the tension, even when life is tossing him a nasty dose of reality.

Listening instead of just responding diffuses tension

Sometimes choosing to connect is more about focus than time. "Jacob was under the weather last week, so I kept him home," Paul volunteered. "But I had a deadline on a report I was working on and felt pressured to get it done. I was in my office, working, while he played in the family room. He's a little sports commentator, carrying on a monologue about whatever he is doing. Every once in a while, he'd say something to me, and I'd just reply 'Ah ha,' never stopping to really pay attention. I was surprised when he showed up at the door, his shoulders drooping, the enthusiasm gone from his voice. 'Dad,' he asked, 'aren't you coming? You said you'd come.' I realized at that moment, that I hadn't truly been listening to him at all. I stopped what I was doing, went with him, and stayed focused. He was delighted and, as a result, so satisfied that he was willing to play alone for another thirty minutes while I finished my report."

It's easy to be physically present without being mentally and emotionally attuned to your child. Create opportunities to tune in. When you're driving in the car, turn off your cell phone, the radio, and the video games, so that you can converse. When you truly listen, you will slow your child's heart rate and quiet his body.

Don't forget your life partner

As you think about creating moments to connect with your child, don't forget your partner or other significant adults in your life. John Gottman has completed extensive research exploring what makes adult relationships work and how those relationships impact children. What he has discovered is that adults in satisfying relationships are more positive with one another. They listen to each other, share humor, and criticize less. Happy, relaxed adults cultivate contented children who know the world is safe enough for them to sleep.

Once again, it's the little things that can make a difference; the kiss hello or good-bye, the compliment or thank you, the question "How can I help you?" The midday "check-in" telephone call, the decision to put down the newspaper and talk, a meal consumed sitting together, the offer of a glass of water or a cup of coffee all matter, the smile, touch, or story shared are all Mother Nature's sedatives tucked in moments of love.

What you also may not realize is that your children are watching. My husband and I were reminded of this fact by the valentine we received from our then eighteen-year-old daughter. On the front, it read, "Every morning for years, I watched my parents across the table from each other, reading a section of the newspaper, occasionally saying, 'Listen to this,' then sharing a bit of news." Inside, it continued, "One day I realized that's what love was."

Choosing to connect keeps you in the "green zone" of calm energy. This is a preventive strategy. It stops the power struggles, sibling fights, and bedtime battles before they even start, allowing your family to move into the evening in a state of calm tired, ready for sleep.

If it feels as though choosing to connect requires time you don't have, or you realize that you're not certain you can remember what it is like to go through a day without feeling tense, perhaps you'd like to check the pace of your life. Steven Covey taught us to make time during the day for the most important things first. It's a lesson we all struggle with, because there are so many things to do.

Slow down the pace of your day

Taking the "rush" out of your day allows you to keep tense energy in check. "It's really hard to slow down," Eve, a small petite woman who radiated energy and enthusiasm for life, began. "Over the past couple of weeks, many of my friends and I have been trying to plan spring sports, summer camps, vacations, etc., for our families. We want to optimize the opportunities we give our kids, but we keep running into the battle of how to balance opportunity with overscheduling. For example, one of my son's friends will have soccer practice from six o'clock to seven o'clock p.m., followed by baseball from seven o'clock to eight o'clock p.m. (at a different park), every Thursday night, for the last month and a half of school. I really want to give my boys, who are six and eight, a chance to play organized baseball and soccer, which they love. But it isn't just one practice. Baseball has two, and then there's a game for each sport. Some of the games are scheduled on Saturdays, but with both boys playing—well, it just feels crazy.

"Together, we made the very difficult decision to forgo baseball and just play soccer, since it is their favorite sport. It wasn't easy. They know their friends will be playing both, but we just couldn't afford to be coming home from practice at eight-thirty p.m. on a school night to shower and get ready for bed. Friday mornings would be misery, and we'd have no time together as a family."

Stop and think before you agree to a schedule of activities so hectic that it takes the "fun" out of them. If you are always running "on the edge," there's no "cushion" for the times when you are caught off-guard. And even if everything "fits" into the schedule, it's still important to consider whether the stimulation and the pace will turn your child into a monster by nighttime.

Watch carefully, and your child will show you if the pace is too hectic. When someone calls and invites you to the park, stop and observe your child for a moment. If she is focused, you will know that she can handle the transition and the activity. If, however, she is bouncing from one activity to another or running around aimlessly, the odds are that she is too tired to handle another outing. This may seem counter-intuitive. Your first thought may be that she appears bored and needs more action, but look carefully. Does she need more stimulation or does she need time to relax?

And listen to your own body as well. When you awaken in the middle of the night, unable to go back to sleep because you are mentally going through your "to-do list," recognize it as a sign of "overload." And when you are missing your partner and friends because you have refused to allow yourself to stop and "chat," too pressured by the need to get to the next event, it's time to review your calendar.

When you choose to slow down, it often feels like a sacrifice. There's always something else to do, whether it's dinner out, a great television show to watch, or an event to attend. Easing tension often means not doing some of these things. What's essential to remember is that after choosing to slow the pace and ease the tension, your child is ready for sound sleep. It's during sleep that a body does its repair work. As a result, it's sleep that actually allows you to do more, because it

- Enhances performance and allows new material to be stored in long-term memory
- Keeps your child healthy by strengthening the immune system, repairing the body, reducing the risk of Type 2 diabetes and childhood obesity
- Allows you to work together more effectively, because well-rested parents and kids find much more joy in each other's company

By talking about tension throughout the day and choosing to manage it, you will eliminate the two-hour fights with your child as he struggles to unwind. You'll have done your work almost imperceptibly throughout the day.

MANAGING TENSION IN THE EVENING

Creating a "nest" for sleep

When I am conducting workshops across the continent, I ask my "host" or hostess to reserve a room for me. I never know quite what to expect. But on this particular night, I was thrilled to enter a historic hotel. A huge chandelier hung in the lobby, telegraphing the luxury of it. Opening the door to my room, I discovered that the bed had been piled high with pillows. A robe was laid across it, and slippers set on the floor next to the bed. Soft music was playing. The drapes had been drawn, and the covers on the bed pulled back. A dim, soft light shone, illuminating cream-colored walls; the entire space was serene, quiet, soothing, and inviting. Despite the fact that I was wound up from the evening's presentation and being more than a thousand miles from home, I couldn't wait to crawl into that bed. When I did, I slept deeply all night long.

We are mammals, and mammals need a "nest," a haven where we feel safe and secure enough to sleep. It's important that you work with your child to establish a "nest." Step into the space where you expect your child to sleep. Look, listen, feel, and smell. Does this environment clearly communicate that this is a place to "shut down" and switch to sleep, or is it an entertainment center filled with interesting toys, an entire library of books, a television, or a computer, begging to be used? Take note of the colors: do they soothe or excite? Is there a mural that dances and stimulates an imaginative mind? Are their mobiles casting shadows or catching your attention?

Now shut off the lights and look again. Rooms look different in the dark. Do you feel safe in this room, or are there shadows that move and frighten you? What light creeps in around the blinds? While the lights are off, listen. Are the sounds you hear comforting or alerting? What is the temperature of the room? If it's too warm, your child will be uncomfortable during the night. Take a deep breath; what are the smells in the room? We now know that smell is actually a very important aspect of an environment for sleep. If it doesn't smell pleasant, or there are hints of smoke, your child's brain will tell him to stay awake.

If you can, lie down on the bed. If your child is sleeping in a crib, place your head at the level of the mattress. What do the crib rails look like from this angle? Is the mattress comfortable? Does the bed feel like a cozy, safe "nest," or does it float in the room, leaving you feeling exposed and vulnerable?

Sheila grimaced as I described creating a "nest." "I'm an artist, and I spent hours creating a very elaborate mural of my son's favorite characters on his wall. I used primary colors. The background was dark blue, the characters' clothes yellow, red, white, green, and black. The curtains were a cotton print that the street light shone right through. He loved to play in there, but he never slept in there. He slept in my room, often on the floor. My room is painted a cream color. Everything else is soft white, with dark shades on the windows. I made it that way because that's what I need to sleep. I realized that he must feel the same way as I do, but I couldn't bear to paint over the mural. So I made his room a playroom and created a nest for him in a little alcove. He loves it."

Keep your child's sleeping space simple, safe, and serene. Remove the television and the computer; put them in a room where it's easy for you to monitor their use. And if you have no other spaces to use, go in before you put your child to bed and together "close it down" for the night.

Once you've created a serene, safe environment, go back through and add the props that help to signal it's time for sleep. Take time to reflect upon the things you love or need for sleep. Watch your child carefully; does she sleep soundly when covered with a heavy quilt or soft comforter? Is she attracted to silky blankets? Include those things that help settle your child for sleep.

Touch eases tension

Studies demonstrate that warm, nurturing touch lowers blood pressure, decreases the heart rate, and reduces stress hormones in the body. When your child asks you to hold him, he's not just being clingy. He's asking you to help him calm his body. You can use this knowledge to help him sleep.

Research studies have supported the positive effects of massage therapy on the health and development of infants, children, and adolescents. In a study of children and adolescents, the massaged subjects were less anxious and more cooperative than the controls. They also had lower levels of stress hormones in their body and slept more at night. After a twenty-minute massage, children were also found to have a more positive mood, be more cooperative, and have lower activity levels. One of the cumulative effects of repeated massage was that children fell asleep more quickly.

What I've discovered in my work with families is that when tension levels are still high at bedtime, fifteen minutes of massage dramatically relaxes the children and eases the "switch" to sleep.

Don't worry if you are not a masseuse, or that it might feel a little strange for your child to ask for a massage. When you recognize your child needs a massage to relax, offer to make a batch of brownies on his back. Begin by pretending to "crack" an egg. Using your fingertips, let the "egg" slide across his back, moving your hands, one after the other, from the upper back to the lower back. Work with your child on the type of "touch" he prefers. Some children like the "tickle," others need a very firm almost "pinch." Sprinkle in the flour. Add the oil and water, allowing your fingertips to flow down across his back. Mix everything together with circular motions from head to lower back, along but not touching the spine. Scrunch up a few cookie crumbs and drop them into the mix, kneading them into his shoulder muscles. Spray the "pan," stroking all along the length of his back. Then spread out the dough, making sure to get it into the corners. Move your hands from side to side across his back.

Once it's baked, it's time to "frost" it, your hands sliding once more across his back. You can end your massage with messages of love by using your fingertip to write the things your child needs to hear: "Good night, Sam. Mom and Dad love you." "Sleep tight all night long!" These are messages that connect and calm, so that he's ready for sweet dreams.

If your child isn't a "back-massage kid"—although you'll find the more you do this, the more comfortable he'll become with touch—start with his arms. Firmly slide your hands down from his shoulder to his fingertips, pulling the stress from his body. Tell him that you are taking out the bubbles, letting them flow out through his fingers. Then do the same with his legs and toes.

When you first start offering your child a massage, you may worry that you're starting a "bad habit." So if you like, you may intersperse massage nights with relaxation nights. On those nights, teach your child to tighten the muscles in his neck and then release them.

Continue down his body to shoulders, chest, arms, hands, belly, buns, legs, feet, and finally toes. He can do this exercise on his own, no matter where he is.

But in my experience, when the day has been filled with calm energy, children don't need extra massage at night. However, when the day has been hectic and tense, as so frequently occurs, a few minutes of massage allows you and your child to end the day with a feeling of connection and calm. You don't have to worry that he'll expect it again at two o'clock a.m., because the odds are he'll be so relaxed that he won't awaken. The key is to stop the massage as you feel his body completely relax and before he actually falls asleep. When you do it, he'll allow you to step away as he slips into peaceful sleep. That's what Kelly discovered.

"We were running behind, and I wanted to shorten the routine, but I hung in there. I turned out the lights and slowly started brownies on Ryan's back. He got a little creative, and wanted me to add extra-thick icing and sprinkles with hearts and stars. I resisted my natural tendency to tell him to stop pushing it, and complied. It took maybe another ninety seconds. He was lying still when I turned on the music and walked out the door, explaining that I'd check back in five minutes. I left his room at 8:38. At 8:40, he was asleep."

Consciously think about your child's touch "diet." If you have a choice between carrying your infant in a plastic seat or your arms, choose your arms. Recognize that when you wean your child, or he begins to use a toilet instead of diapers, significant "touching" is lost to him. Replace those intimate times. Hold your child on your lap while you read to him or brush his hair. Offer to "scratch" an older child's back. Your touch sends a powerful signal to your child's brain, letting it know he's safe. It is all right to sleep.

Questions to end your day with

In order to sleep soundly and fall asleep quickly, your child's body must be calm. You have the opportunity to set the stage for success throughout the day. When you stop, look at your child and listen attentively, you'll make him feel treasured and loved. Through your touch, tone of voice, and action, you help him to relax his body. When you reduce the sense of urgency in your day, you bring your child into a state of tranquility, ready for sleep. So, tonight, when you tuck her into bed, take a moment to reflect upon your day. Even during those times when life seems determined to take a battering ram to your door, ask yourself a few simple questions: Did you stop and connect, or did your child have to fight to stay awake in order to have time with you? Did you slow down enough for your family to sleep? Change begins with awareness. If today you had to answer "no," tomorrow will offer you another opportunity. You can make a different choice.

Managing the tension prepares your child for sound sleep. It is not, however, the only factor. Setting your child's body clock is also crucial. That's why it is so important to recognize the little things that can innocently disrupt it.

EIGHT

TIME

Recognizing what upsets your child's body clock

"We cannot control the wind, but we can set the sails."

—Elisabeth Kübler-Ross

Homemade chocolate fudge sauce beckoned to me. I'd made it Sunday night for dessert, slowly melting chocolate chips, butter and sugar, and then pouring it, still warm, over French vanilla ice cream. Two days later, I knew the leftover sauce was still in the refrigerator. Bedtime loomed. I was tired, ready for sleep, but I craved that sweet, chocolate richness. I could reheat it in the microwave and pour it over the ice cream, letting it harden slightly so that it would cling to the cold, smooth cream as it melted in my mouth. I teetered, knowing that I was making a choice. On Sunday night, I'd succumbed to my cravings and snatched a small second serving as a bedtime snack. An hour later, I was tossing and turning in bed, unable to fall asleep. The magic of the chocolate's caffeine combined with the excitement of my day struck, leaving me wide-eyed and awake. Yet tonight I stood in front of the refrigerator, poised to gamble. Perhaps this time it wouldn't bother me. Sometimes it didn't. The day had been more relaxing. I was very tired, ready for sleep. What difference could a little hot-fudge sauce make? Surprisingly, it can matter a great deal.

ALL DAY LONG WE ARE MAKING DECISIONS THAT AFFECT OUR SLEEP

Ironically, it is often the little decisions made throughout the day that can be setting you up for power struggles. Think about it. Have you ever

- Allowed your child to have a Coca-Cola with lunch or a Mountain Dew with dinner?
- Let your child skip his nap?
- Held a slumber party for your child and her friends?
- Allowed your child to watch television, "chat online," or check e-mail in the evening?
- Attended an athletic or social event with your child that went past his normal bedtime?
- Roughhoused with your child right before bedtime?
- Discovered your child has fallen asleep in the car while you were running errands after four o'clock p.m.?
- Offered your child the opportunity to stay up late as a "treat"?

If, like me, you answered yes to any of these questions, you may unknowingly be upsetting your child's body clock.

How the body clock works

The body clock is the control center for the sleep/wake cycle. It tells the body to be awake during the day and permits sleep by turning off at night. What researchers have discovered is that the sleep/wake cycle, or what researchers like to call the circadian rhythm, runs on a cycle closer to twenty-five hours than twenty-four. In order to bring your child's cycle into line with a twenty-four hour day, you have to set it with cues, like light and a regular sleep-and-wake schedule. If you don't set it, your child's body clock will tell her to be alert at the wrong time, leaving her wide awake when you want her to be sleeping, and drowsy when she is supposed to be awake.

You make a difference

The decisions you make throughout the day can inadvertently be confusing your child's body clock, throwing it out of balance and, as a result, making it more difficult to fall asleep and to sleep soundly. That's what happened in Emily's family.

After reviewing the list of questions, Emily shook her head. "I had to say yes to almost every one of them! Our extended family went on vacation together. We were having such a great time; we just kept going, skipping naps, and staying up late. When Hannah wanted a soda pop, Grandpa bought it for her. She was pretty good at least the first few days, but since we got home, it's been pure agony for two weeks. She won't stay in her own bed. She can't fall asleep, and she's so whiny and irritable." She shook her head and groaned.

A good night's sleep doesn't just happen. It requires managing the tension, so the body is relaxed enough to sleep, as we have discussed, AND setting the body clock, so that the brain knows when to "switch" to sleep. The key is to be thinking about sleep and, as a result, making decisions that set the stage for it by cueing the brain when it's time to be awake and when it's time to sleep. By identifying the little things that innocently disrupt the body clock, you can choose to avoid them or to plan around them. By doing so, you allow Mother Nature to become your ally. She's a powerful force, and with her on your side, bedtime struggles disappear, because you are working with your child's natural body clock instead of against it.

You may be innocently putting your child into "jet lag"

If you've ever experienced jet lag, you know what it's like to be "off-kilter," working against your body clock instead of with it. Michael flew from Minneapolis to Los Angeles on the 11:20 a.m. flight arriving in LA at 1:20 p.m. Skipping lunch, he dashed to an afternoon meeting followed by a dinner with college friends. The lively conversation renewed his energy and, as a result, it was two o'clock a.m. in Minneapolis by the time he fell into bed.

Hours past his normal bedtime, he lay in the dark, wide awake. Images and ideas streaked through his mind like northern lights on a cold winter night. Pulsating with energy, his body seemed to hum. He rolled to his back and tried taking a few deep breaths. Nothing! Forty-five minutes later, he finally fell asleep, only to awaken again after two hours. He got a drink of water, fluffed his pillow, and tried again. This time, he awoke at his usual Minneapolis morning time. It was four o'clock a.m. in LA. He willed himself to sleep, but it was hopeless. Giving up, he dressed, ate breakfast, and headed for his morning meeting, arriving early. His brain foggy, he fought to stay focused and grew impatient with the discussion, wishing someone would get to the point. But what was the point? He couldn't remember. His eye twitched. He wondered if anyone else noticed, and then, as though from a distance, he heard his name. Startled, he realized he'd been asked a question.

Michael was experiencing jet lag—and two thousand miles away, in Minneapolis, so was his eight-year-old son Brent. But it wasn't a business trip that had done Brent in. It was the "treat" of staying up late on Saturday to watch the playoff game.

Brent's symptoms were not obvious to his mother. Instead, they looked like misbehavior. At bedtime, he complained he couldn't sleep, needed another glass of water, and begged for one more story to be read. He couldn't fall asleep. When he finally did, he woke frequently during the night and notified his mother each time. This was not a ploy to torture her. Like his father, he didn't sleep because he couldn't. It was not a willful action, but a body clock off-kilter.

Changing habits isn't easy—but it's worth it

If you are feeling skeptical, thinking this is just one more thing to think about or that it may require a schedule that feels too rigid or is unrealistic for today's families, you're not alone. Nicole admitted, "I heard you say, 'Sleep is a choice and it has to be a priority.' I still didn't think it would work. But I remember feeling overwhelmed. Our daughter Joanna was consuming our lives, rolling from one meltdown to the other twenty-four hours a day. Thinking about the decisions we were making was just one more burden. She was already taking all of our time, but we were desperate and willing to try anything. In the end, we gained so much more freedom. Once we became aware of the things that disrupted her clock and stopped doing them, she was able to sleep. Her behavior became more predictable and steady. It made it possible to do things with her we would never have tried before."

It's true that changing habits isn't easy, but it can really be worth it. So let's take a look at what's happening in your family. Go ahead and identify the innocent decisions that may be inadvertently disrupting your child's body clock. Then you can decide whether or not changes are warranted and worth the effort. It's always your choice.

Take a look at what's happening in your family

At the end of this chapter I have included a weeklong sleep journal for you and your child. It is VERY important for you to collect this information. It will be used to help you decipher your child's natural patterns and recognize the decisions that are playing havoc with your child's body clock. But since I suspect you may not want to put this book down for a week while you gather it, I'm going to provide you with a shortcut. Take a few minutes and complete the following chart.

	Weekday	**Saturday**
When did your child go to bed?	________	________
When did your child fall asleep?	________	________
When did your child get up?	________	________
When was breakfast?	________	________
When was lunch?	________	________
When was dinner?	________	________
How much time did your child spend watching television, or on the computer?	________	________
When did your child watch television or play on the computer?	________	________
How much exercise did your child get?	________	________
When did your child get exercise or roughhouse?	________	________
When did your child nap?	________	________
When did your child consume beverages or foods containing caffeine, i.e., soft drinks, hot chocolate, chocolate desserts, candy, ice cream, coffee-flavored desserts, etc.?	________	________
How much time did your child spend "restrained" in chairs, car seats, grocery carts, strollers, etc.?	________	________
When did your child play outside?	________	________

Now use this chart to check yourself as we go through the next section.

INNOCENTLY DISRUPTING SLEEP

In a study conducted by Dr. A. Kahn, researchers found that 58 percent of children's sleeplessness was attributed to little decisions. So what are these little things that matter so much? There are four main categories:

1. Irregular schedule
2. Light
3. Lack of exercise or exercise at the wrong time
4. Stimulants

Together they create the "**ILLS**" of the night, playing tricks on your child's body clock, and your own. As you review them, think about each individual in your family. How sensitive your child is to the disruptions depends upon her temperament and current tension level. One child may be highly susceptible to even the slightest alterations, while another seems to go more easily with the flow. And when tension levels are high, changes that upset the body clock are even more detrimental, because the body is already on alert. High-tension levels combined with a body clock out of equilibrium can make bedtime an event to dread hours before it arrives. So, review each of the following culprits to identify those that impact your child most significantly.

Irregular schedule

Wake times When it comes to synchronizing your child's sleep/wake cycle, the very first decision of your day may matter the most. If the time your child gets up keeps changing, it may be the real reason behind those conflicts. Dr. James Mass states, "You must resynchronize your internal clock every morning. Otherwise your natural 'alarm clock' for alertness will be buzzing at the wrong time in your sleep/wake cycle."

Looking at the chart you just completed, check what's happening in your family.

	Yes	No
Is there more than a thirty- to sixty-minute difference between when your child awoke on the weekday and on the weekend?		

"Wait a minute!" Jamie declared. Folding her arms across her chest and jutting her chin out sharply, she demanded: "Are you saying that we're not supposed to sleep in on the weekend? That's one of the few luxuries I still have in my life! Don't even think about taking that one away!"

"Remember my story about craving a hot-fudge sundae?" I asked her. She nodded. I smiled, as I admitted: "I opened that refrigerator door and ate every last morsel!"

Jamie sat back, at least for the moment willing to continue.

It's not that we NEVER do these things, or create a schedule so rigid that there's no room for flexibility or fun. Our goal is to become aware that we are making a choice that can be disrupting sleep. Recognizing the decisions that interfere with the body's natural clock allows us to choose whether or not it is worth the cost. Then, when we knowingly make a choice to do something that will likely disrupt our child's sleep, we can make accommodations for it by investing a few extra minutes in helping him to wind down, or simply extending an extra dose of patience at bedtime. We can also adjust our routines the day before and/or after. Knowledge gives us power.

When I realized what I'd done by eating the hot fudge and the caffeine it contained, my first thought was, I couldn't give up chocolate! What I've come to realize is that I can still have my hot fudge. I simply won't choose to have it before bedtime—at least most of the time, and certainly not on a night before a big speaking engagement.

So why would you even consider limiting how long or how frequently you sleep in? It's tempting to think that if your child goes to bed later, you'll simply let him sleep longer. But a surprising thing happens. If your child is young, and especially if he typically awakens at a specific time, it's unlikely that he will sleep an extra hour. More likely, he'll awaken at his normal time—an hour short of sleep and ready to fight with you. By naptime, he'll be overtired. Sleep deprivation creates tension, which activates stress hormones, putting his body on alert. As a result, he'll be unable to sleep until later in the afternoon, when he finally "crashes" at four o'clock p.m. and sleeps until five-thirty p.m. Because of the late nap, it's very likely that he'll be up watching the ten o'clock news with you. The next day, you'll be awakening him again, because he went to bed late. He'll be short on sleep, and the erratic schedule will continue. His internal time-keeper will struggle to function without a clear point to guide the "switch" from awake to sleep.

Even for infants, a regular wake time in the morning clearly communicates that this is the time our family gets up and gets going. It is one of the first cues to help him differentiate daytime from nighttime.

If, by chance, your child does sleep later, as most adolescents will do, take note. The sleep is unlikely to be of the same quality. He will awaken groggy, take longer to get going, and still—despite the extra sleep—feel fatigued. That's because he's not in sync with his body rhythm. While there will be times you want to let your adolescent sleep in, and, in fact, need to let him, when you are aware of the body clock, you can work together to reduce the amount of divergence in his schedule and minimize the impact of an erratic schedule. Paul remembered, "My mom always insisted that I get up by ten o'clock a.m. on the weekends, instead of noon, like my friends did. I knew I'd have to get up, so I tended to head home just a bit earlier. I think it made it easier to get going on Monday morning."

Check your morning wake times. Are they fluctuating frequently, causing your child to start the day out of sync?

Irregular bedtimes Every night you are making another decision that plays a fundamental role in how easily and how well your child sleeps. Shifting bedtimes can be throwing your child into "jet lag."

	Yes	No
Is there more than a thirty- to sixty-minute difference between when your child went to bed on the weekday and on the weekend?		

"We've figured this one out on our own," Terri told the others. "Over the weekend, we were out with friends on Saturday night, and let the kids stay up late, as a treat. Sunday night, they got to bed an hour past their usual bedtime, too. On Monday, they were miserable, completely inflexible, and so emotional that they were falling apart over anything. Yet they couldn't settle down to sleep when I put them to bed that night."

"But I always wanted my kids to be flexible," Jenna lamented as she listened to Terri. "I've intentionally moved around their bedtimes so they wouldn't be rigid." Then she sighed, as she admitted, "My older two are so easygoing and low-key that they've survived it, but my youngest, he can't handle it and has never slept well. It's almost funny; I've tried to make him more flexible when actually I've made him extra inflexible because he's so exhausted."

There is a "window" for sleep, when the body clock tells the brain to "switch" from alert to sleep. Sometimes that window can be very narrow. When you move your child's bedtime even fifteen to sixty minutes, you may discover that instead of sleeping better because he's really tired, he can't fall asleep and wakes more frequently.

This doesn't mean that you'll never spend an evening with friends, let your child stay up later on a weekend night, or go to pick up Mom or Dad at the airport after a long business trip. You'll just think about it before you do it, knowing that it may make tomorrow a tougher day.

Irregular meal times

Yes No

Is there more than a thirty- to sixty-minute difference between when your child ate breakfast on the weekday and on the weekend?

Yes No

Is there more than a thirty- to sixty-minute difference between when your child ate lunch on the weekday and on the weekend?

Yes No

Is there more than a thirty- to sixty-minute difference between when your child ate dinner/supper on the weekday and on the weekend?

Breakfast, lunch, dinner, and snacks all play a role in setting the body clock. If serving time fluctuates, the body has no clear cues that this is the time to be active and taking in nutritional resources.

Irregular meals also lead to hunger. Hunger creates tension. If your child is not receiving on a predictable basis the food that he needs, tension increases, making it even more difficult to sleep.

There's also a natural dip in energy right after lunch. If lunchtime varies, without meaning to you may miss the natural "window" for a nap and have set the stage for a meltdown rather than a siesta.

It's true, there are children who never seem to fall into any type of predictable schedule. What you'll discover is that, despite their resistance to it, these children actually benefit greatly when the world around them helps to set their body clock.

Know your child. Avoid attempting to force him to eat when he's not hungry—a conflict that only adds to everyone's tension level. But do take note of meal and snack times and what happens to your child's sleep and behavior when the schedule fluctuates.

Irregular nap schedule or skipping naps Whenever a child under five years of age goes more than eight hours without sleep, the body pushes harder to stay alert. The natural late-afternoon dip in energy is overridden, and the body clock is thrown off.

Yes No

Is there more than a thirty- to sixty-minute difference between when your child napped on the weekday and on the weekend?

Yes No

If your child is five years old or younger, does she go more than 8 hours without a nap?

Yes No

Did your child nap after three o'clock p.m.?

"I'm not much of a schedule person," Christine admitted as she read through these questions." I've heard too many people say they can't go to the park because it's two o'clock p.m. and it's naptime. Why can't their kids just nap at three o'clock, or skip it one day? Is it really going to matter that much? I've got friends who are constantly checking their watches and then rushing home to ensure their kids get their naps. Where's the spontaneity?"

Others quickly joined in. Robert declared, "My kids nap at daycare, but they never nap at home." And Sheila insisted, "Sam gave up his nap the day he turned two. It wasn't worth the forty-five-minute battle trying to get him down."

"If my daughter does take a nap, I wake her up," Paul stated, "otherwise she's up until midnight." And Lynn added, "I've got three kids under five. We would never get out of the house if I waited for everyone to get their naps."

In unison, they glared at me, waiting for a response. I ducked before replying!

Every child is different, but be aware that the decision to delay a nap, skip it completely, or let your child catch a catnap later in the day may be setting you up for bedtime battles.

On the days your child misses his nap, does he wake more frequently at night? Many children do. And if you, like Robert, have a child who naps at school but not at home, go back to your chart. Check the weekday column. What time did he get up in the morning? When were breakfast and lunch? When did he go down for a nap? Now look at what happens on the weekend. Do you sleep in—just a little bit? Was breakfast more of a brunch? Did lunch get delayed until early afternoon? Did you try to put him down for a nap thirty minutes or more later than he experienced during the week?

Many children are thrown off by that shift in their schedule—even if it's minimal. The time change can mean the difference between sweetly snuggling up and going to sleep, and forty-five minutes of fighting over a nap that never does occur. Or at least not until hours later, when he crashes in the car for twenty minutes, or on the couch—just long enough to get a second wind, but not long enough to be truly restored. His body clock is thrown off, and he's left short on sleep.

And then there's the fear that if he does nap, he won't go to bed at a reasonable hour, but once again, look at what time he had his nap—was it completed by three o'clock p.m.? If not, the nap is probably occurring too late. (Unless you are a "night owl" family, and late rising and bedtimes work for you.) Check also what time you tried to put him down at night. Often, because a child has napped, it seems logical to keep him up later, but the bedtime may then be too late, the window for sleep missed.

Now go back and take a look at your overall schedule. Does it allow Mother Nature to work with you? When your child complains he can't fall asleep, there's a reason, and it may be an irregular schedule.

It's hard to believe that fifteen to sixty minutes could really matter, but they could. A shift in the schedule only one or two nights a week means that 29 percent of the time your child's sleep is disrupted. Think about it. One decision that YOU have the power to make has the potential of cutting the tantrums by nearly a third.

LIGHT

Morning light

Light, strong morning light, is one of the most powerful influences on the body clock. It is the cue that this is the time to be awake. Dr. Chuck Czeisler has discovered that dim light, such as electric room lights or the light from a television screen, has a much more subtle effect but can also reset the body clock.

	Yes	No
Did your child spend time outside *only* in the afternoon and/or evening?		

It's easy to let the kids wake up in the morning, flip on the television, and lounge on the couch, but that decision can be unwittingly creating havoc with their body clock and stoking the fire for meltdowns later in the day. That's because the brain isn't getting a clear signal that THIS is the time to be awake.

If your child is a night owl by nature, his first inclination will not be to get up and get going. While still respecting his need for a slower start, know that limited exposure to morning light leaves the body clock in limbo.

Light at the wrong time including screen time: television, DVDs, computer and video games

What's more likely to disrupt your child's sleep:

a) A brother sharing the room

b) A television in the bedroom

c) A fan in the bedroom

The answer is b) a television in the bedroom. What could television or screen time have to do with your child's body clock? Exposure to light helps to set the body clock, but it can also disrupt it. If you are saving your child's favorite video for a bedtime treat, you may be innocently making it more difficult for him to fall asleep or stay asleep. If your child is very sensitive and watches television or instant-messages with his friends close to bedtime, the light from the screen can trick his body into thinking it's early in the day, instead of time for bed. In fact, if your child is highly sensitive, more than an hour of screen time, any time during the day, may be playing games with his body clock.

Yes No

Did your child have more than an hour of screen time, i.e., television, video, video games, computer time?

Yes No

Did your child experience screen time after six o'clock p.m.?

Yes No

Does your child's room receive light from the setting sun or a street light?

Jerking forward in his chair, John snorted: "That's ridiculous. By the time my kids finish their homework, they've spent hours on the computer. There's no way we could limit screen time to an hour a day!"

A recent study completed by the Kaiser Family Foundation found that children are spending an average of six and a half hours a day in front of a screen, either working or playing. Extended screen time is a reality today. But if your child is misbehaving and experiencing difficulty sleeping, it's important to look at the number of hours spent in front of the screen and when it is occurring. Some children will be impacted more than others. Recognizing screen time as a potential culprit behind your daily struggles allows you to set limits on game time, and to schedule homework completion as early as possible in the day. The younger and more sensitive your child, the greater the likelihood he'll be affected.

"That's what must have happened!" Robert exclaimed. "I like to unwind by playing games on the computer," he said. "Last night, I held the baby in my lap while I played solitaire. Afterwards, it was so much harder to get him down. He's so little. He's not playing the game. I never imagined that the light would be an issue for him."

Paul looked sheepish. When I turned to him, he said, "It was my turn to put the kids to bed. I let them watch a video at seven o'clock. My wife told me not to, but I thought what the heck. It took forty-five minutes longer to get them to bed than the nights they don't watch television. The next morning, they were crabby at breakfast."

Bright light is a good thing in the morning, but at night it can trick the body clock. That's also why it's often difficult to simply come home from the shopping center, a gymnasium, or a hockey rink, and hop into bed. And why on summer nights, if you live in a northern climate and it's light until late at night, your child doesn't want to go to bed. It's not a power play when your child tells you he can't sleep. It's his body clock telling him: time to be awake.

Check the lights—are you using them to your advantage by letting in enough sunlight in the morning? Or is exposure to light at night, including electric lights and screen lights, tricking your family into thinking it's time to be awake?

LACK OF EXERCISE, OR EXERCISE AT THE WRONG TIME

Physical activity clearly designates wake time and creates healthy fatigue that promotes deep, sound restorative sleep. Recently, the National Academy of Sciences suggested that adults and adolescents need at least thirty minutes of physical activity a day. Younger children need twice that much. But, on average, the teens and adults surveyed said they exercised thirty minutes a day only four times a week. And, according to an article by psychologist Robert Brooks, less than 25 percent of school-age children get even twenty minutes of rigorous daily physical activity.

Yes No

Did your child spend more than an hour today restrained or strapped into car seats, strollers, etc.?

Yes No

Did your child have less than an hour of physical activity today?

Terri's eyes opened wide, as she exclaimed, "I had no idea how much time Katie spends strapped in. Katie is the youngest of my four. Every morning, when she wakes up, I strap her into her high chair for breakfast. As soon as she's finished, I pop her into her car seat to take the older kids to school. Then, since I only have her, I usually do my shopping. Katie goes into the shopping cart—strapped in again. By the time we finish, it's usually time to pick up my son from preschool, so Katie goes back into her car seat and stays there until Ben comes out of school. Sometimes she falls asleep in the car for a few minutes. When we arrive home, they're both famished, so I put her back into her high chair and feed her. Until I completed my sleep chart, I didn't realize she'd been up four and a half hours and basically strapped in the entire time! She never had a chance to move. No wonder she fights her naps so badly on those days."

Of course, if your child is in a car or high chair, she needs to be restrained. But what you may have not realized is that because of your "efficient" scheduling of all the errands together, your toddler may be spending half of her wake time immobilized. Lack of exercise undermines sound sleep.

"That's so true," Amy said. "My four-year-old twins have to have their exercise. We try to get it in the morning, but if we miss it, I make sure they get out in the afternoon. Otherwise, by dinner, they are out of control, not listening, wrestling, and unable to settle down."

Is your child getting the exercise he needs to experience healthy fatigue and clearly communicate to his body clock that this is the time to be awake?

Exercise at the wrong time

While exercise at the right time promotes sleep, exercise at the wrong time of the day may leave your child too energized and hot to sleep. One of the signals for the body to "switch" to sleep is a natural drop in body temperature. Physical activity raises the body temperature. That's why roughhousing or exercising too close to bedtime may rev up your kids instead of wearing them out.

"Oh ho, talk to my husband about this one!" Kelly declared, poking an elbow toward Paul sitting next to her. "I've asked him not to do it, because they get all wound up and won't go to sleep. But he still does it." She glared at Paul who huffed in frustration.

"My kids love it when I get down on the floor with them. That's what dads do!"

Cameron jumped in, agreeing with Paul. "No roughhousing before bedtime? I can't avoid it at my house. Not with three sons. I'm not going to yell at them for being boys. I'd have to completely separate them. They're couch jumpers. Running around and wrestling is what they love to do. Usually, I do make them stop twenty minutes before bedtime, and I have to massage my middle son to settle him down, but the other two—when it's eight o'clock p.m. they're out, always have been."

"Everyone's right," I declared. Looking at me expectantly, they waited for me to explain.

Cameron's sons provide a perfect example of individual differences. It appears that two of the boys are so "regular" that no matter what they are doing, their clock will "switch" to sleep. However, for the middle one, too much activity—especially roughhousing—may throw him off his natural rhythm and require a bit of assistance from Mom to help him get back on track. So, both Cameron and Kelly are correct: some kids can tolerate it, and some kids can't.

And Paul's right, too: many kids do love to roughhouse with their parents. I have wonderful memories of crawling on my own dad's back for a horseback ride before bed. Just be aware that these activities can be arousing rather than soothing. CHOOSING the timing so that physical activity doesn't interfere with the body clock is essential. Think about the pre-bedtime activities at your house: are they helping your child's body to slow down and "switch" to sleep, or are they raising the temperature indicating it's time to be awake? If you have a child who has trouble falling asleep after roughhousing or playing hard outside before bed, a shift in the timing can make all the difference.

STIMULANTS

Caffeine

It may be difficult to believe that the treat of a cola or Mountain Dew with pizza at lunch can be the reason your child won't go to bed at eight o'clock p.m., but once again, this simple decision may be impacting your child's behavior.

Caffeine is a stimulant. It increases activity in many parts of the brain, delays onset of sleep, shortens overall sleep time, and reduces the depth of sleep. It also increases frequency of urination and can cause tension, anxiety, sweating, and elevated blood pressure. Even a breast-fed infant experiences the rush of caffeine ingested by his mother, which leads to irritability, increased heart and pulse rates, and altered blood-sugar levels.

A low dose of caffeine is considered to be 80 milligrams a day for adults. That's one cup of drip-brewed coffee, or two colas. The effect of caffeine is related to body size and weight. When a child drinks one can of a cola beverage, the effects of the caffeine are comparable to an adult drinking four cups of coffee. And the less frequently a child has caffeine the more it affects him.

Caffeine in beverage form reaches all tissues of the body within five minutes of ingestion. Peak levels are reached in thirty minutes. Half of it is metabolized in four hours, less rapidly in young children. Approximately 50 percent of the caffeine consumed at three o'clock p.m. is still in your child's body at seven o'clock p.m.

Caffeine is often included in beverages, foods, and medications where you might not expect it. Anacin and Excedrin tablets contain as much caffeine as some coffees, iced teas, hot teas, and soft drinks. Dr Pepper, Barq's Root Beer, Mountain Dew, and Sunkist Orange, as well as caffeinated waters, coffee-flavored ice creams and yogurts, chocolate candy and desserts, and energy drinks all contain caffeine. As a result, when your child has a few sips of soda, a couple of bites of a chocolate dessert, and one energy drink, you may not be aware of how much caffeine she is ingesting.

Every child is different, but observe closely and read labels. Look at the days your child has difficulty falling asleep and staying asleep. Is caffeine sneaking in his bloodstream, sending his body clock into a spin?

Intent and result

Now go back through this chapter and review each of the culprits. Which ones do you suspect may be playing tricks with your child's body clock?

The number of things he had been doing that might be upsetting his son's sleep/wake cycle struck Paul. "He was always a flexible kid," he said. "So we just took him wherever we were going. We never thought about naps or wake times or bedtimes or meals. Unfortunately, this flexibility was wearing Jacob out. He was a walking zombie, losing it over the tiniest things during the day. It felt so out of our control, but now I realize that it's not. We really can make a difference."

What's most exciting is that it's YOUR decisions—the things you DO have control of—that can truly matter. You can stop and ask yourself two simple questions: (1) what is our intent? and (2) what will be the likely result? If you are skipping a nap in order to have fun as a family, it's important to recognize you just made a decision. What will be the result? Will everyone enjoy themselves, or will you end up with overtired and cranky children for the rest of the night and perhaps the next day? Sometimes it's worth the pain. Sometimes it's not. You can decide.

Simply by observing closely so that you know which decisions are most important for your child, you can help set your child's body clock. By doing so, you reduce the time it takes for your child to fall asleep AND improve the quality of his sleep and his behavior.

One more incentive for change

In case you need one more enticement to begin making changes for your family, researchers have discovered that the things that interfere with sleep can also stifle the adult libido. So, if you'd like to improve things in that department, too, take note of your decisions.

SLEEP JOURNAL

Name ______________________

Age __________

DAY	5 a.m.	6 a.m.	7 a.m.	8 a.m.	9 a.m.	10 a.m.	11 a.m.	12 p.m.	1 p.m.	2 p.m.	3 p.m.	4 p.m.	5 p.m.	6 p.m.	7 p.m.	8 p.m.	9 p.m.	10 p.m.	11 p.m.	12 a.m.	1 a.m.	2 a.m.	3 a.m.	4 a.m.	Total Sleep Hrs
Sample		WA		M TV	TV			M	QT				M	O	BT	ST									10
Sun.																									
Mon.																									
Tues.																									
Wed.																									
Thu.																									
Fri.																									
Sat.																									

Behavior Notes:

Sunday: ______________________

Monday: ______________________

Tuesday: ______________________

Wednesday: ______________________

Thursday: ______________________

Friday: ______________________

Saturday: ______________________

Key: ***For Sleeping***

Shade areas when sleeping
SW = Spontaneously wakes
WA = Woke by alarm/other
BT = Bedtime

For Eating

M = Meal or Snack
C = Caffeine

For Awake Time

TV/ C = screen time
O = Outside time
A = Active time
QT = Quiet time
R = Movement restricted (car seat)

NINE

A GOOD NIGHT'S SLEEP BEGINS IN THE MORNING

Setting the body clock

"When it comes to the development of your child, quality wake time is not more important than quality sleep time."

—Claire Novosad, Southern Connecticut State University

The ice disappeared from Lake Minnetonka on April 6 at 3:45 a.m. this year. In Minnesota, any sign of spring is eagerly sought. The event is closely monitored; the earliest "ice out" took place on March 11, 1878, the latest, May 8, 1956. That's a spread of fifty-nine days between the earliest and the latest. Mother Nature can be a teaser. The most common ice-out dates are April 17 and April 18. On those two dates, the ice has gone out nine times. The average of all of the dates is April 15. Ironically, the ice has never actually vanished on this day.

I tell you all of this because Mother Nature not only plays with ice-out dates; she also plays with body clocks. There's a wide breadth of what is considered normal, and it's rare that anyone actually hits the average. I have to use averages, however, in attempting to help you set your child's body clock. Your job then is to tweak the recommendations to fit your child and your family. If your child is an infant under six months of age, his body clock is working differently. That's why I have a chapter specifically for infants. But if your child is older than six months, you can begin to "nudge" him in a way that helps him set his clock and live within the bounds of your family's schedule.

Routine versus rut

Routines create predictability in our lives. They're efficient, comfortable, and convenient. And studies demonstrate that individuals with a regular schedule report fewer sleep problems and more resistance to depression.

What we wish to avoid, however, are routines that become so monotonous or rigid that we lose all spontaneity and creativity in our lives. So what I intend to provide you with are practical strategies for establishing a consistent schedule and clear cues that tell your child's body when it's time to sleep and when it's time to be awake. Don't worry; I won't give you one perfect schedule. Instead, I'll describe schedules that actually work in a real world, where Thursday's events are totally different from those of Monday, and don't even come close to what happens on Saturday.

Nor will I offer an "ideal" schedule that fits one child, when the reality is that you have three. You'll get to design the final product for your family. And if, by chance, the mere mention of the word "schedule" evokes a deep sense of irritation and rebellion within your soul, or a groan is spontaneously erupting from your throat, stay with me. I promise to not only teach you how to create a schedule, but how to play tricks with it when the desire hits. Most important, I'll show you how the benefits can outweigh the effort.

Envision sleep

The schedule that we are about to create may look different from those you've established in the past. Frequently, schedules evolve as we review our responsibilities and commitments, marginalizing sleep as something we do in our spare time. But Steven Covey tells us that when climbing the ladder to success, we have to be sure our ladder is lying against the right wall. We need a vision of where we are going in the long run. In this case, the vision I am leading you toward is one of calm energy, peak performance, and joy. Our "ladder" lies against the wall of "choosing sleep," understanding that adequate, sound sleep brings enormous benefits to our lives. That means the FIRST thing that goes on the schedule is sleep.

If your child is of school age, you are going to budget ten hours for sleep, then school time, and other significant events your family values the most, like worship, family time, or community service. Only after these essentials are in place, will you add other activities. If there isn't room for all of your child's interests, it will provide you with a wonderful opportunity to teach him about making choices in order to establish balance in one's life.

Here are the average sleep needs for each age group:

Age of individual	**Average hours of sleep needed over 24 hours**
Infant 0–12 months	14–18 hours
Toddler 13–36 months	13 hours (including nap)
Preschooler 37–60 months	12 hours (including nap)
School-age 6–12 years	10–11
Adolescent 13–19 years	9.25
Adult 20 years	8.25

I want to stress once again that these are averages, which means that 50 percent of individuals need slightly more and 50 percent need less sleep. In my experience working with families, I have found that when given the opportunity, children tend to sleep LONGER, often up to an hour more than the average. You'll know you and your children are getting enough sleep when you are awakening on your own before the alarm goes off, feeling energetic and ready to face the day. (If your child snores and is getting the average amount of sleep for his age group, yet is still fatigued, it may be the first indication of a medical condition or sleep disorder. See your doctor.)

Terri sighed deeply as I wrote the averages on the board. "With our commutes and jobs, we're not even home long enough for the kids to get that much sleep."

Together, we brainstormed potential solutions. They included: Grandma picking up the children and starting dinner for the family, Mom and Dad staggering schedules allowing one to go to work early and pick up the children in the afternoon, while the other went to work later and handled the morning routine; extending afternoon naps so that the children needed fewer hours of sleep at night, or sending "dinner" to daycare so that the kids were ready to begin the bedtime routine immediately upon arrival home. Finally, we explored the rather drastic solutions of changing jobs in order to limit work hours or moving to reduce the commute time. Sometimes making sleep a priority merely requires creative tweaking. Sometimes it may lead to a significant life change. You will get to decide. But once sleep is identified as a priority, your goal will be clear, and you'll discover there are many creative options for achieving it.

A good night's sleep begins in the morning

When Diana met with me, her goal was to get her two-year-old daughter Lydia on a schedule. Diana never quite knew when Lydia would awaken in the morning. Sometimes it was as early as seven-thirty, but other days she would sleep until nine o'clock. As a result, her nap was also unpredictable, often not occurring at all. Bedtime was even more challenging, gradually drifting later and later until it hit eleven o'clock p.m. Exhausted, Diana was determined to make a change, but she didn't know where to begin.

You can wake your child

While you can't MAKE your child sleep, you can create an environment that encourages sleep and WAKE her up. A good night's sleep actually begins in the morning, when a morning wake time is established.

I asked Diana to keep a sleep journal for a week, like the one in the last chapter. From it, we discovered that Dad had a flexible job and liked to spend time with Lydia in the morning. But on those days that she slept until nine o'clock, he didn't start work until nine-thirty or ten o'clock, and then began his day feeling rushed. He also missed dinner, because he had to stay late to complete his work.

Together, Diana and her husband decided that it worked best for their family when Lydia woke at seven-thirty. The sleep journal showed this was definitely within Lydia's "normal range," so I readily agreed. Two weeks later, Diana reported: "I started making sure Lydia woke by seven-thirty. If she was asleep, I'd open the blinds, start to sing to her, or pick her up. We let her come in our bed and play with us, and then we'd all get up. By creating a predictable wake time, I also set her naptime and bedtime. Now she's taking a nap at eleven-thirty, going to sleep at eight o'clock every night, and waking on her own at seven-thirty a.m. My husband has time to have breakfast with us and is still out the door by eight-thirty a.m. Because he's going to work earlier, he's able to come home for dinner. It's wonderful."

The question, of course, is how do you know when wake time should be? Turn to the sleep chart found at the end of the previous chapter. Use the information you collected there to complete the following summary form.

What's the pattern?

Day of the week	Mon.	Tues.	Wed.	Thurs.	Fri.	Sat.	Sun.
Wake time	____	____	____	____	___	____	____
Had to be woken y/n	____	____	____	____	___	___	____
Breakfast	____	____	____	____	___	___	____
Lunch	____	____	____	____	___	___	____
Dinner	____	____	____	____	___	___	____
Nap time*	____	____	____	____	___	___	____
Nap time*	____	____	____	____	___	___	____
Bedtime*	____	____	____	____	___	___	____
Time actually fell asleep	____	____	____	____	___	___	____
Exercise	____	____	____	____	___	___	____
TV/screen time	____	____	____	____	___	___	____
Outside time	____	____	____	____	___	___	____
Total hours of sleep	____	____	____	____	___	___	____

**Note whether your child fell asleep E (easily) or D (with difficulty).*

Identify your child's goal wake time

Review the wake times for the entire week. What's the range? What's the average? What's the earliest your child either woke or had to be awakened? At what time did your child awaken of his own volition? Finally, was this a typical pattern or was there something quite unusual about the week? If this was an unusual week, keep a sleep journal for at least one more.

The Petersons were both employed full-time. Getting out of the house in the morning without screaming was their goal. They often found themselves literally pulling their three-year-old son Nathan kicking and crying out of his bed. When the Petersons and I reviewed the chart for Nathan, it looked like this:

Day of the week	**Mon.**	**Tues.**	**Wed.**	**Thurs.**	**Fri.**	**Sat.**	**Sun.**
Wake time	7:00	6:00	7:00	6:00	7:00	8:00	8:00
Had to be woken	Yes	Yes	Yes	Yes	Yes	No	No

The range of wake times for Nathan was six o'clock a.m. to eight o'clock a.m. The average was seven o'clock a.m. The earliest Nathan had to be woken was six o'clock a.m. It was only on the weekends, when he could sleep until eight o'clock a.m., that Nathan awoke on his own. Unfortunately, his preferred wake time of eight o'clock a.m. did not fit with his family's lifestyle. One alternative was to make the average wake time of seven o'clock a.m. the goal, but that left Nathan short on sleep two days a week when his parents had to go to work early. Those two early mornings held the potential of creating a cycle of sleep deprivation that set him up for trouble. So the goal morning wake time for Nathan became six o'clock a.m. From this, we were able to set his nap-, meal-, and bedtime, ensuring that he got enough sleep and ultimately began waking independently at that time. When he awoke on his own, he was ready to get up and get going, and the morning power struggles disappeared almost entirely.

The decision to make six o'clock a.m. the goal wake time didn't necessarily mean that Nathan and his parents would never be able to sleep in. It was true that if Nathan was a "regular" child, we could predict that he would eventually awaken at six o'clock a.m., even on the weekends. However, this wake time also left open the option of sleeping in extra minutes if the routine had been disrupted a bit. Thus six o'clock a.m. as the goal gave the family room for flexibility and spontaneity. Now, if bedtime runs a little later on Tuesday, Nathan has the option of grabbing up to another hour of sleep Wednesday morning if he needs it. The family isn't running on the edge of sleep deprivation with no room for flexibility. (Occasionally sleeping in for less than an hour won't disrupt the clock for most individuals—although you'll need to observe your child closely in order to discover how much leeway you have. And when you and your child are getting adequate sleep there's no need to "sleep in.")

Now select your goal morning wake time by identifying the earliest hour your child has to arise. Then stop and ask yourself, Is it early enough to allow getting ready without feeling rushed? Is there sufficient time to "switch" from asleep to awake gradually? When you and your child have time to lie in bed for a few minutes, collect a few cuddles, listen to a little music, and then start moving, the day will begin more smoothly. Once you've decided, note it on the "Goal Schedule" included at the end of this chapter.

Adjust for the "morning lark"

If your child is an early riser and waking earlier than you'd prefer, select a morning wake time that fits your family better. I'll show you in chapter 16 how to move his wake time. However, if he's a "morning lark" type of person, there is a genetic factor underlying his early awakenings. As a result, if he's awakening at six o'clock, it's unlikely that you'll be able to move his wake time to ten o'clock a.m. You can get him to seven o'clock, and potentially eight o'clock, but after that, you're pushing it—until he reaches puberty. Then the hormones of puberty will turn him into a night owl, and you'll remember fondly the days when he popped out of bed so easily in the morning.

Adjust for "night owls"

There is also a genetic factor underlying in children who are night owls. The kids who like to stay up late at night and sleep in the morning have a specific variation of the CLOCK gene, and, as a result, a slightly altered circadian rhythm. An early-morning schedule runs counter to their natural body clock. You'll have to be even more consistent about maintaining the morning wake time in order to keep them on a schedule. Their natural tendency will be to immediately flip out of it and return to later bedtimes and wake times.

Rather than fight with the night owl's natural body clock, if you have a choice between two schools, one starting at seven-thirty a.m. and another at nine o'clock a.m., choose the latter. It will fit better with the night owl's preferred cycle. This is especially true for the adolescent. During puberty, researchers have found, the circadian rhythm shifts. Even the child who was a morning lark as a youngster suddenly finds it difficult to fall asleep at an early hour and even more challenging to get up in the morning. If your adolescent is attending an early-start school, you will need to be more flexible about sleeping in on the weekend to allow him to get enough sleep. Or, if you are of the mind to make social change, the Edina Minnesota Public Schools moved start time from seven-thirty a.m. to eight-thirty a.m. The number of citations for disruptive behavior, and the absentee rate dropped drastically, while test scores rose.

Once you've selected your goal wake time, write it down on the "Goal Schedule" at the end of this chapter.

Determine your child's sleep time

Now that you've selected the morning wake time, the next step is to determine sleep time. Sleep time is NOT bedtime. Sleep time is the moment when your child is asleep. Bedtime is the time you begin your routine of preparing for bed and the final wind-down before sleep time. Thus, sleep time may be eight-thirty p.m., while bedtime may be eight o'clock p.m. for a child who needs thirty minutes to get ready for bed and unwind before falling asleep. Another child may need an hour. His bedtime, therefore, would be seven-thirty p.m., for a sleep time of eight-thirty p.m.

Frequently, bedtime and sleep time are determined according to when Dad and Mom get home for supper, or when a favorite television show ends. And, then again, it might depend on when a book is finished, or how well the kids have been behaving; the better the behavior the later the bedtime. The fact is that bedtimes and sleep times are often determined not by a child's cues of fatigue but by how they fit in after all the other activities of the day. In our plan for setting the body clock, sleep time and bedtime are determined by focusing on how much sleep your child really needs.

Check the averages

Look at the average sleep chart presented earlier in this chapter. If your child is napping during the day, subtract that time from the total. (If he's five years old or younger, I'm going to strongly encourage you to reintroduce a naptime, but we'll discuss that later.) Your child's goal sleep time is determined by moving back from his goal wake time. For example, Nathan's goal wake time is six o'clock a.m. Preschoolers need an average of twelve hours of sleep in a twenty-four-hour period. Nathan naps two hours every afternoon. That means he needs ten hours of sleep during the night. His goal sleep time is then eight o'clock p.m. If he didn't nap, his sleep time would be six o'clock p.m. If he was a school-age child who required an average of ten hours of sleep, his goal sleep time would also be eight o'clock p.m.

Adjust for your child

If you review your child's sleep journal and realize that he's sleeping far less than the average for his age (i.e., a preschooler sleeping eight-and-a-half or nine hours a night), it's very likely that he is severely sleep-deprived. He's so tired that he can't sleep. So, rather than assuming that he needs less sleep than his peers, use the average sleep need for his age group as your guide to determine his goal sleep time. This gives you a place to start. You'll fine-tune it as you work with your child.

Be prepared to adjust your goal sleep time for the season as well. Susan Perry writes, "In the fall, we are like our fur-bearing, hibernating mammalian peers. We have a desire to store a little extra fat and have a growing desire to sleep away the impending winter."

So, if in June your preschooler requires only eleven hours of sleep, don't be surprised if she needs to add another hour of sleep beginning late fall, especially if your November is dark and dreary, and a cold winter is breathing down your neck. We are smart mammals, and our bodies know, if it's nasty out, we really should spend more time nested comfortably under the covers.

The average sleep chart is just that, an average. It helps you to identify the ballpark you're working within, but it needs to be fine-tuned to fit your child. If you watch and listen closely, your child will ultimately show you when you've found his true sleep time by falling asleep faster, sleeping more, and behaving much better. The key is to look for his "window" for sleep.

Catching the "window" for sleep

It had been a long trip to Grandma's, and dinner was with unfamiliar relatives, but one-year-old Eric was thoroughly enjoying his birthday party. He tore away the paper of his presents, especially intrigued by one: a coffee can covered with contact paper. It had a slit cut into the top, and orange-juice can tops to drop into it. Crouching, he worked hard to slide each lid into the can while unopened presents and new toys lay strewn about his feet.

Suddenly the clock struck eight o'clock. He rubbed his eyes, then stumbled and fell. He sought his mother for comfort, rubbing his eyes again as she held him. Once he was comforted, she set him down. He walked back to his coffee can; his steps slow, as though drugged. He stumbled and fell again. His mother picked him up, said, "Tell everyone night, night," took him to the bedroom, put on his pajamas, nursed him, and laid him down, covering him with his little blanket. He went to sleep. His astute mother had recognized his "window" for sleep. Even though it was not convenient for her, and the presents were not all opened, she stopped and put him to bed. Her reward was an evening spent chatting with other adults and a happy toddler the next day.

Catching the "window" allows your child to slip into sleep

Your child's "window" for sleep is the moment at which Mother Nature is telling him, time to "switch" to sleep. When you recognize it and respond accordingly, your child will fall asleep within minutes. There are some children who make it easy for you. Their urge to sleep is so strong they will tell you that they are tired. Other children, like Michael, won't say it, at least not yet, but demonstrate cues that are very easy to identify. Some children's signals are much more subtle. If this is the case for your child, you'll have to watch and listen carefully, but the clues are there, although often for only fifteen to twenty minutes. For those children, when you miss the "window," their body will immediately kick into a "second wind." It'll be at least another forty-five minutes to an hour and a half before the next "window" for sleep will appear, and then, because they are overtired, it's likely that they will have more difficulty shifting to sleep.

When the "window" for sleep is difficult to identify

Review your child's sleep journal to see if there is a consistent time when he's quickly falling asleep for his nap and at night. If no clear pattern emerges, you can identify your child's "windows" for sleep time and naptime more easily if you set aside a day or two and stay home, simply for the purpose of finding them. Plan to avoid running errands or going out. Instead, organize activities at home, in the yard, or in the park nearby. Then watch and listen. If your child has experienced difficulty falling asleep in the past, the cues will likely appear an hour to an hour and half earlier than you might expect. So, what do the cues look like? Here's what other parents have told me:

"Elizabeth is a cheerful baby. When she's in her saucer, she gurgles and plays happily, but if she's fatigued she growls and snorts. If you pick her up and walk past her crib, she'll actually reach out for it."

"Tommy becomes 'fragile'; little things that earlier in the day wouldn't have bothered him upset him."

"Sarah wants to cuddle. Her body droops; she yawns and rubs her eyes."

"Eight-month-old Becca begins to rub her head or simply puts her head down and starts to hum. Frequently she holds a soft toy to her face."

"Six-year-old Christopher can't make a decision."

"When Joey, the high energy kid, hits his 'window,' there's a momentary slowing of his body. He'll stop and just sit, maybe take a car or several and stay in one place playing with them instead of running around or crawling on the floor. He might also choose a book or blocks, something quiet, instead of a noisy toy like a fire engine. Other times, he'll simply come and lean against me. But it's only for a second, and then he takes off again."

"Eight-year-old Angela's eyes glaze slightly and her voice loses its energy."

"Fatigue becomes apparent when Tyler starts looking for his blanket and pacifier."

"If I'm rocking Lisa, she'll get to a point when she doesn't want to be held. She'll arch her back. If I put her down, she'll fall asleep. I can even tell the babysitter: when she arches her back, she's ready for sleep."

"Todd complains that he's bored (and starts looking for sweets)."

"Eddie crawls up into the La-Z-Boy chair and lays his head down on the arm of it."

Whether it's a blatant yawn, a drooping of the eyelids, a "zoning out," a change in your child's breathing, or a sudden lack of coordination, your child will show you her window. The moment you see the signals, start preparing your child for sleep. Or, if you're like Ellen, you may need to be even more observant. "I find I have to have my three boys ready for bed BEFORE the first cues appear. If I wait for them, they move into overtired before I can get all three of them down. But if they're all ready for bed, the moment I see that pause, I pop them into bed and they go right out."

The challenge is when the "window" appears at an inopportune time. That's what Robert saw. "There's actually a 'window' for our kids right around six-thirty, but we're still on the road at five o'clock and just sitting down to eat at six-thirty. There have been times we've skipped dinner and put them to bed."

Changing your routine or adding a nap can help you adjust your child's "window" to fit your family's schedule. But sometimes the family has to adjust a bit for the child as well. Wendy dropped a morning class when she realized it fell right in the middle of her son's "window" for nap. "It's only for a short time," she told me. "Soon he'll grow out of his morning nap, and I can take the class again, but for now this is what he needs."

You'll know when you find your child's "window" for sleep, because it will be so much easier for him to fall asleep. Eyes sparkling with excitement, Latisha reported, "Miguel's window is seven o'clock p.m. If I put him down, then he goes right to sleep. But last night, we were running late and I didn't get him to bed until seven-forty-five. It took him much longer to fall asleep. I never thought that a few minutes could make a difference, but they definitely do."

By identifying the "window," you get Mother Nature on your side, helping to move your child toward sleep. Sometimes, however, when you find the "window," your child may still not be inclined to head for bed. In that case, you'll have to be the one providing the gentle nudge that moves him toward the sleep he needs. It may require making hard choices and isn't always convenient, but if you keep in mind the ultimate goal of sound sleep, I suspect you'll find it easier to follow through.

Respond before your child moves into overtired

Sometimes, despite your best efforts or intentions, you may miss your child's "window." You'll know it, because he will move into overtired behaviors. "I didn't realize it," Jamie said, "but I was waiting until Katie was overtired. She gets crabbier and crabbier, until she actually growls." When your child is crying, has gone into a frenzy of activity, becomes aggressive or silly, starts talking back, has over-the-top reactions, or can't be satisfied no matter what you do, he's moved past his "window." Or, if he's shrieking and streaking through the house, becoming more and more hyper, a second burst of energy has grabbed him and propelled him on into the night. If this is the case, note the time and move your goal sleep time fifteen minutes earlier the next night. Continue doing so each day until you catch the "window" and she falls asleep easily.

It's not just the children who have a "window" for sleep. Adults do, too. And taking note of your own and honoring it instead of pushing through to complete one more job, or one more phone call, can magically stop the conflicts with your child—because you, too, will be getting the rest you need.

Establish bedtime

Bedtime is the time you need to begin helping your child get into his pajamas, have a bedtime snack, brush his teeth, use the bathroom, take a bath, select books, read, and, if needed, get that last wind-down massage, or moments of reflecting on the day. You will want to allow enough time for the routine, so that you do not feel rushed. Otherwise, you will be inadvertently creating tension when you are trying to help your child be relaxed enough for sleep. Some children can move through this transition quickly, and only need thirty minutes. Others may need an hour to an hour and a half. Check your child's sleep journal to see what time you've started the bedtime routine, and when it's worked the best. You can also use your own needs as a guide. Are you a person who stops what you're doing and literally jumps into bed? Or do you require time to check the doors, wash your face, put away your clothes, or complete any other last-minute job before you can allow yourself to stop? If you or your child's other parent are individuals who need more wind-down time, it's likely that your child does, too.

Parents are often surprised at how early their child's bedtime needs to be in order for him to get the sleep he truly needs. "Before we took the class, we never thought of putting him to bed at six o'clock," Sheila reported. "Sam doesn't nap and frequently wants to eat at four-thirty or five o'clock. In the past, I'd give him juice or something to tide him over, but now I just feed him dinner. I moved his bath to the morning. When his dad comes home at five-forty-five, he starts to read to him, and at six o'clock he's asleep. Initially, he wanted to get up at four-thirty a.m. and was awakening a few times a night. But I just told him it wasn't time to get up yet. Now he isn't waking at all until six o'clock a.m."

As you adjust your child's bedtime to fit his true sleep needs, it may end up being so early that it leaves you little to no time with him in the evening. If you've been together all day, that might be just fine. The early sleep time allows you to have adult time before you need to go to bed. But if you've been apart all day, or have a partner who works late hours, without a nap you'll be forced to choose between letting your child go to sleep when she needs to, or seeing her parent. It's a very tough decision. Naps can give you more flexibility. So, let's go back and add naptime to your goal schedule, even if your child isn't presently napping.

Schedule naptime

Naptime and the number of naps during the day that your child needs are going to vary according to her age and her "style." Infants may require three or more a day. Older infants, eight to nine months old, may drop to two a day. Toddlers up to eighteen or nineteen months may still require two, while some have dropped to one long afternoon nap by a year. Toddlers are likely to be ready for their naps earlier than preschoolers. Except for the infants who can take a late-afternoon nap and still go to sleep at a reasonable time at night, the preference is to have naps completed by three o'clock p.m. Unless, of course, your family enjoys having little ones up during the evening. Once again, it's that combination of reading your child's cues and working with him to fit within your family.

If your child is an infant, I'm going to refer you to chapter 15. If you have a young toddler, it's very likely that she's ready for her first nap about one and a half to two and a half hours after she wakes up in the morning. If she awoke at six o'clock, sometime between seven-thirty and eight-thirty a.m. she's ready to go down again. Check your child's sleep chart to see if she fell asleep during this time, especially on a low-key day. If she didn't, begin watching for the telltale signs: rubbing the eyes, slowing her pace. I suspect you'll see it right about this time. Set your goal naptime to give you a starting point. You'll tweak it as needed. Once your child is on a schedule and things are going smoothly, it's likely that she'll sleep about ninety minutes to three hours. If she sleeps less, that's fine, as long as you can see that she's rested.

Spacing naps

The space between the first and second nap is often a bit longer, but, once again, depends on the child. Some children are ready to go down again two hours later, while others will be up for three. So, if your one-year-old naps from eight until nine-thirty a.m., she'll be ready for her second nap between eleven-thirty and twelve-thirty. Again, the second nap may range from ninety minutes to three hours in length. Remember, these are averages. You have to observe and chart your own child to discover her rhythms. Unfortunately, you won't just have to do this once. As she grows and develops, her sleep needs will change, requiring you to adjust accordingly. Our motto will remain "progress, not perfection."

If your child is a preschooler, after lunch is the opportune time for a nap. If your child is attending a program that allows him to nap during the week, match that time on the weekend for your schedule. Now, go ahead and add nap times to your goal schedule.

Meals make a difference

Meals and regular snack times also help to tell the brain when it should be alert and when it should shift to sleep. Schedule breakfast shortly after your child awakens. Even if he's not a breakfast eater, offer something light, to cue the brain that it's time to start the day. Without that cue, the schedule can easily start to drift. Add lunch. If your child eats lunch at school or child care at eleven-thirty, you'll want to follow that same routine on the days he's at home. Finally, add dinner. I often find that dinner is served too late for young children. If your child is complaining about the food or completely losing it because you poured his milk in the blue cup instead of the yellow one, it's very likely that he is too tired to eat. If this is the case, consider feeding him between four-thirty and five-thirty and then allowing him to have a quick bedtime snack with the late-returning adult. This schedule permits you to still have a family meal, while keeping your child on a schedule that allows him to get the sleep he needs. Researchers suggest that it's best if we don't eat a heavy meal right before bedtime. So, if you can, schedule your child's meal at least an hour and preferably two before sleep time. Remember, too, to eliminate the caffeine from your child's diet.

Now check your summary chart to find the most consistent mealtimes that work well for your family. If they fit with your child's sleep needs, continue serving at that time. If they don't, tweak the schedule, making sure that the alterations work for you. Fit healthy snacks in between the meals.

Exposure to light matters

Get up and get out Morning light clearly communicates to the brain that it's time to be AWAKE! That's why it's so important to get your child up in the morning and outside. If you can, include in your schedule time for your children to play outside before heading to school. And don't forget the babies. They, too, benefit from a morning stroll, or a walk with the dog. Add exposure to morning light to your goal chart, to cue your child's brain that this time, rather than two o'clock a.m., is the time to be awake.

Limit television and computer time

There are three reasons to limit the television and computer time for your child. One is the exposure to light, which can trick your child's brain into thinking it's the middle of the day. The second is that your child will likely fight to stay up and watch the end of the show. And third, the stimulation, even if it's only on in the background, can put your child in a state of alert. The issue, of course, is what do you do when you turn off the television? My family had to figure that out the day our television set died.

My children had just finished third and sixth grade the summer that our television set broke. Summers are leaner on a teacher's salary, and since we were a one–TV set family, its demise meant we were "unplugged."

I noticed first what we stopped—namely, fighting. Tussling over the remote, arguing about what show would be watched, debating how many shows could be viewed, and begging, mostly on my part, to turn it off in order to get dressed, come to dinner, or go to bed. It was an unexpected relief to discover how easily it was to transition from one part of the day to another.

I didn't miss the quarrels, but what I did miss was the diversion late in the afternoon. Television had provided amusement while I cooked dinner. Now, at the end of the day, it was unavailable. The children seemed determined to fill the void by picking on one another. Between the clatter of pans were the shouts of "That's mine!" Or "That's not fair!" I was forced to come up with another alternative, so we began cooking together, searching through my cookbooks and magazines, finding recipes to keep them interested. Kristina discovered she had a talent for baking. Joshua, perhaps because he was studying French, took an exotic flair and began creating French cuisine for us. I wonder now if it was that summer he decided to become fluent in French and live and study in France.

Initially, the summer days did not start well. The kids did not know how to begin their day without the click of the television set. "We're bored!" they'd complain, and so, once again, I was forced to think.

This was not easy. I was used to my mornings spent with the newspaper and a cup of coffee. Unused board games came off the shelves; paints, markers, and chalk that lay dusty in their boxes appeared on the table. Tools tossed onto the workbench in the garage were sorted and organized, offering an invitation to use them.

Suddenly, instead of getting up to loll in front of the television set for an hour or two, they were up, dressed, and out the door, working on projects. One morning, a neighbor called. In a tentative voice, she asked, "Do you know that Kristina is cruising the neighborhood, taking plastic two-liter soda bottles out of the recycling bins?" I was not surprised. She and her friend Kellen were determined to create a raft that would float on the pond behind Kellen's house. They collected two hundred two-liter plastic bottles and duct-taped them together. We all went to the "launching" and watched as they pushed it off into the murky water. We saw it float for ten seconds and then slowly sink below the surface, leaving them both with slimy thighs and grins on their faces. I wonder if it was that summer that Kristina decided to become an engineer.

Books from the library were also selected. We began reading together, never a favorite activity for Kristina, so we started "family read," taking turns reading aloud to one another, stories like *Hatchet* and *Call of the Wild,* tales of adventure that intrigued us all. Perhaps that's why they later both chose to go to college in the mountains, and why, at the end of the summer, Kristina had jumped five reading levels.

In the evenings, without television, there were basketball games in the driveway, with the previously unused basketball hoop. We went on bike rides, concerts at the park, and, when we needed a "fix," a movie, albeit at times the children sat a few rows down, with their friends. But we saw it together, and could talk about it afterwards.

I could tell you that we remained "unplugged," but that would not be true. Minnesota winters are long, with short days and long, cold nights. We bought a new television set, not a big screen, just a little one that provided our "fix." But by then, the habit was broken. We never did turn it back on in the mornings, nor was it an automatic response when the kids were bored.

Television viewing and computer games can play havoc with your child's body clock. Review your summary chart. If your child has more than an hour of screen time a day, it may be affecting his sleep. If you would like to change your child's habits, consider going "cold turkey" for a few weeks, during which you completely shut off the television and eliminate computer games. Many of the families I work with actually find this approach easier than trying to gradually cut back. If this seems a much too drastic strategy, start cutting back. You can choose the method that works for you. One easy step may be to simply get the television and computer out of the bedroom. Or you might try consciously selecting the shows you want to watch and the games you want to play, then turning off the equipment. Most important, avoid watching television or playing computer games right before bed. The result will be better and more sleep. The extra bonus will be your children's enhanced reading skills—a proven benefit.

There were grumblings in the group the night we talked about turning off the television. I never expected John to come to my aid, but he did, with flying colors. "Mary came to our home," he told the group. "We hadn't had a decent night's sleep since our three-year-old son Ethan was born. She convinced us to start waking him, establish a schedule, and turn the television off. We did it for two weeks, and the results were amazing. He slept better, stopped whining, and the number of tantrums really dropped. But I couldn't believe that television could really matter that much."

He paused, cocking his eyebrow and glancing in my direction. "So, I decided to experiment. There was a football game on, and I let him watch it with me. He went to sleep easily, but he awoke five times during the night, and, once, it took nearly an hour to get him back to sleep again. I still wasn't convinced. So, the next night we watched his favorite video together. That night, I had a great night's sleep and the next morning I awoke feeling triumphant—until I saw the look on my wife's face. I'd slept through it, but she'd been up with Ethan most of the night." He turned back toward me, shaking his head. "Now I'm a believer. I even recommended it to another guy at work yesterday."

If the thought of turning off the television set turns you off, go ahead and experiment. Some children are not as sensitive to it as others. It's all right to find out what your child needs, but I suspect that, like John, you'll discover the sacrifice is really worth it.

Dim lights in the evening

Mother Nature created sunsets to cue us that the day was ending. Thanks to the invention of electrical lights, we've been able to ignore her cue. But if you'd like your child to fall asleep more easily, begin dimming lights in the evening. Put a dimmer on your dining-room light. Use a night-light in the bathroom instead of turning on the bright overhead lights. The kids will love reading by flashlight or a soft bedside lamp. During summer, when darkness may come very late and morning light very early, use dark-out blinds to help your child's body know it's time to "switch" to sleep.

Schedule exercise

Exercise provides a double benefit. When it comes to setting the body clock, exercise at the right time of the day clearly tells the brain this is the time to be awake, and tires the body in a healthy manner. It carries the additional bonus of reducing tension and, if done on a regular basis, toughens the exerciser, making him more resilient and better able to handle stress. So, making sure your child gets enough exercise during the day is extremely important. Not only will it help him sleep better, but, as Micah's parents discovered, it also kept him in bed. "If Micah doesn't get enough exercise and outside time, we have to pick him up and return him to his bed at least five or six times before he'll stay there. But if we make sure that he gets enough exercise during the day, the curtain calls disappear."

Exercise can start even with infants. When you bathe your baby, allow enough time for him to kick against the water. Instead of letting him sit in a plastic carrier that confines him, place him on a blanket on the floor. If there are carpeted stairs, put the safety gate up on the third and let him practice crawling.

Encourage older children to get outside and move. When sixteen-year-old Brent complained that he couldn't sleep, his mother encouraged him to start running two miles a day. After only ten days, he told her, "I can't believe how much better I'm sleeping." Eight-year-old William added time to ride his bike before getting on the bus in the morning. Not only did he sleep better, but reprimands for not sitting still in school disappeared.

But what if you live in Minnesota, where it's fourteen below zero on a January morning, or in Arizona or Florida, where it's just too hot on a summer afternoon to get outside. I asked parents for suggestions for bad-weather days.

"We head for the mall. I park far away from the entrance, so that we have farther to walk. They know they have to hold my hand, and do. Inside, there's a ramp they love to run up and down, and I just let them go."

"Dancing is one of our favorite pastimes. We have tile flooring in our dining room, so we just push the table to the side. Everyone has tap shoes and really gets into it. Yesterday, we were dancing to a Barbie tape—even my husband. It was a hoot."

Angie is known for her creativity, and it shows in the activities she thinks of for her kids. "We each choose an animal, and then run around the living room or crawl up and down the steps, acting like it. Sometimes, I lay out pillows as stepping stones, pretending the carpet is water, and the kids must jump from one stone to the other or they'll get 'wet.' If I want them to be quieter, I hand them a roll of electric tape and a basket of little cars. They 'lay out roads' with the tape. They'll spend at least an hour crawling across the floor, driving the cars down their roads."

The key is to be consciously thinking about opportunities for physical activity on a regular basis. Search garage sales or toy stores to find equipment that encourages movement, such as a small trampoline, an exercise ball, or an indoor jungle gym. Be sure to add "heavy" work, like digging, pulling a wagon, wheelbarrowing on one's hands across the room, carrying books upstairs, or walking uphill. Exercise that includes repetitive motion, such as swinging, walking, running, or bicycling, is especially soothing and calming.

Just remember that exercise tends to raise the body temperature, so it's recommended that the activity be completed at least three hours before you want your child to fall asleep. Once again, this is an average—some individuals can work out and head for bed, relaxed and ready for sleep, but the activity leaves others energized. Select the best time for your child to get physical exercise. Add it to your goal schedule. If you can, combine that activity with exposure to daylight, especially morning light.

Select other activities

Your goal schedule now includes all the key elements to set your body clock. Now it's time to add in other activities. These are the things that bring joy to your life. You can see that if some of your favorite things to do are active, you can have fun and be setting the body clock. Other choices may intrude upon your child's sleep time.

The challenge in all of this is that we don't live alone. It's not just one child's schedule that we are trying to create, but an entire family system that needs to be addressed. That's when one of my most favorite slogans comes into play: "We're a Problem-solving Family." People often laugh when I share this slogan with them, but it's a very powerful statement. It immediately connotes to your children that in this family, we consider everyone's needs and work together. And that's what it takes to create a family schedule that meets the needs of all.

It's not easy to make sleep a priority. Especially when you realize that your child's bedtime falls smack in the middle of your favorite "Dad and Me" music class, or that your toddler's nap should begin when his older sister is supposed to be in a swimming class. If the event is only one time a week, go ahead and experiment. Make all of the other schedule changes for setting the body clock and see if that one night or one afternoon really throws things off or not. If it does, you may consider registering for another time slot, for example, Saturday morning instead of Wednesday night. Or, if the timing is close between the end of the event and the beginning of your child's sleep time, you may choose to adjust your child's overall schedule slightly to accommodate that activity. You can also establish carpools or even hire a sitter to let one child sleep while you're out with another. And then again, you may look at the openings on your goal schedule and realize that for right now the best thing for your child is to not select that particular activity. The choice is always yours. The key is to be thinking about sleep, planning for it and making it a priority.

Ben and Dana made sleep a priority. "We decided to get a bit tougher about when we put our kids down at night. It's hard in the summer, when everyone else is still outside, but if our kids are sleeping, we can always welcome friends and neighbors in our home for some quiet socializing. Putting them to bed, we also found, gives us more time together as a couple rather than two people who simply cohabitate."

Go back to your goal schedule and fill in the free-time slots that work best for your family. Remember, children benefit from downtime as well as active time—and so do you. Once you've made your choices, review your entire goal schedule. The key is to find the times you can consistently honor, thus allowing your child and yourself to set the body clock and work with your circadian rhythm.

The value of a goal

The nice thing about a goal schedule is that it gives you something to work toward. If you know you are aiming for an eight-thirty p.m. bedtime, it's easier to remember. And when you choose to skip signing up for the second sports team, you won't feel guilty, knowing that you are choosing to have a schedule for your child that includes sports AND sleep.

If following a schedule is new to your family, don't get discouraged if it takes two to three weeks for the changes to be effective. You are adjusting the body clock. It moves slowly, but it does change. You'll also notice differences in how each family member adapts to the schedule. That's why it's time to talk about temperament.

GOAL SLEEP/ACTIVITY CHART

Name ______________________ Age __________

For Sleeping: WA = Wake time
N = Naptime
B = Bedtime
S = Sleep time

For Eating: M = Meal or snack

For awake time: E = Exercise or active play
O = Outside time
A = Other activities

TIME	Sample	Monday	Tuesday	Wednesday	Thursday	Friday	Saturday	Sunday
5 a.m.								
6 a.m.								
7 a.m.	WA							
8 a.m.	M							
9 a.m.	O/E							
10 a.m.								
11 a.m.								
12 p.m.	M							
1 p.m.	N							
2 p.m.	N							
3 p.m.								
4 p.m.								
5 p.m.	M							
6 p.m.								
7 p.m.	B							
8 p.m.	S							
9 p.m.								
10 p.m.								
11 p.m.								
12 a.m.								
1 a.m.								
2 a.m.								
3 a.m.								
4 a.m.								
Total Hours of Sleep	12 Hrs							
Behavior								

TEN

TEMPERAMENT

Is your child's genetic "wiring" making it more challenging to sleep?

> *"We tried putting her in the crib and letting her cry, but we couldn't stand it and neither could the cat. So she slept with us or on a little cot next to us."*
>
> —Terry, mother of three

When my son was born, I was already teaching classes for parents. I'd read all of the books and knew what to expect—or so I thought. The problem was that Joshua, my firstborn, had not read the books. Despite what the experts told me, he did not quickly fall into a schedule. Nor would he nurse then calmly fall asleep in his bassinet. No, Joshua was the kid who only fell asleep at the breast or lying on his father's chest, or being held in our arms. Even if we waited until his breathing had slowed, the moment we moved to lay him down, he would awaken, and the whole process had to begin again. It was a humbling and frustrating experience listening to the parents in my classes whose babies could be fed, laid in a crib, and then promptly went to sleep.

In my fog of exhaustion, I wondered what we were doing wrong. I even considered changing careers. How could I call myself a parent educator when I couldn't even get my own son to sleep?

Three years later, my daughter was born. She was the "textbook baby" I'd read about. After nursing her, I expected to rock her, as I had done for her brother. But Kristina squirmed in my arms, clearly communicating that she wanted to be put down. When I did, she stuck her thumb in her mouth and immediately fell asleep. I stood over her crib and wept, fearing that one day she would discover how many hours I'd rocked her brother and be angry because I hadn't rocked her.

Early on, it was apparent that Kristina was a natural-born sleeper.

When she got older, she often asked to be put to bed, and it wasn't infrequent that she missed her bedtime story because she fell asleep so quickly.

How could this be, two children, same biological parents, same house and beds, yet two very different styles? Puzzled and intrigued by their differences, I began to review the research and discovered all is not equal in the land of sleep. Some children and adults find it more challenging to "switch" from alert to sleep. The reason, researchers are beginning to believe, may be in their genes.

What is temperament?

The most recent research demonstrates that there is a genetic influence in how an individual reacts to the world around him. This first and most natural response is called temperament. Some individuals, by their very nature, are more easily stirred up. When surprised or faced with a potential threat, they startle, their heart and pulse race, blood pressure rises, stress hormones elevate, and more activity occurs on the right side of their brain. As a result, it takes more energy and skill for them to moderate their responses and calm themselves. I've written extensively about these spirited children.

Other individuals seem oblivious to the same events; their responses are low-key and much easier to manage. These differences in responsiveness, it is believed, are regulated by the central nervous system. They are not caused by birth order, bottle- or breastfeeding, the difficulty of birth, or where or with whom one sleeps. Rather it appears that children are born with a tendency to act and to react to people and events in their lives in specific ways that can be identified and predicted. The reactions are relatively consistent for each child in different situations and at different times. One result is the variation in how easily children fall asleep.

Today researchers in the fields of psychology, physiology, anthropology, and neurobiology are actively exploring temperamental differences. They agree about the reality of temperament and the important role it plays in children's experiences. But because they represent many fields, they tend to use a variety of names to describe the traits. I choose to use the terms coined by Dr. Stella Chess and the late Dr. Alexander Thomas, because of their parent-friendly approach. They include the *intensity of our reactions*; our *sensitivity* to sights, sounds, smells, lights, textures, smells and/or emotions; our *speed in adjusting to changes*; our *energy level*; and *how regular or predictable our body rhythms* are.

A child who is temperamentally sensitive not only notices all of the sights, sounds, and smells around him, as well as the tension level, but also must sort this information and decide what it means. Telling him to ignore the strange smell of a new detergent on his pillowcase is like telling you to ignore someone pricking you with a needle. He can't do it, even when he wants to—instead, he needs you to help him stop the sensation or, if that is impossible, to teach him coping strategies.

You do make a difference

When Thomas and Chess began their work, the prevailing theories at the time insisted that all children arrived as blank slates. Their personality was determined by the parenting they received. Nurture, not nature, made the difference. But Chess and Thomas, parents themselves, quickly realized that in too many cases the nurture-alone approach did not explain the differences they observed at home or with their patients. Now, as the result of their work and the work of those who have followed them, we know that nurture AND nature matter. How one's temperament is ultimately expressed depends on one's age, experience, and training.

While genes provide the template for how the body will respond, the environment provides the opportunities for learning the skills to manage the responses. By tenderly nudging and practicing, you will increase your child's ability and capacity to cope. By doing so, the researchers now believe, you provide the practice needed to potentially create new pathways within the brain and, as a result, fresh ways of responding and functioning.

You cannot choose your child's temperament, but you can make a big difference. It is you who recognizes that the intense child needs help soothing and calming himself. It is you who provides the gentle wind-down so he can sleep, and who will gradually teach him how to soothe himself. It is a process that takes time, but as you provide the structure and the skills he needs, he will be more relaxed and sleep better. The more he sleeps, the more adaptable and open to your guidance he will be.

Work with your child's temperament instead of against it

Ignoring or denying his temperament only pushes him into a state of tense energy, further from the sleep he needs. Your child's temperament signals what is happening inside of his body and how it is reacting to the world around him. Understanding temperament will help you discover why he struggles to sleep when others don't. It is this awareness that can help you keep your cool, predict the potential trouble spots, and take the steps that will help your child get the sleep he needs.

Until Wyoma, the mother of three, discovered temperament, no one in her family was getting any rest. "I was exhausted," she told me. "But I didn't know what to do. My first two were so easy. I fed them and laid them down, and they went to sleep. I thought I was a great mom—until Tyler came along. I had always heard that the third child was supposed to be easy. Tyler must not have heard the same story. Behind his back, we referred to him as the PITA kid—pain in the a . . ! Anyway, it felt like he was fighting me. Even if I held him, he'd cycle his arms and legs fiercely in sharp, quick movements. I knew he was tired. Sometimes his eyes were shut, yet he wouldn't sleep. I thought he was intentionally tormenting me. And then I learned about temperament.

"All of a sudden, I understood he wasn't fighting me. He was fighting his own body. He's much more intense than my other two. Little things upset him and wind him up, and if he gets overtired, there's no way he can settle himself. When I recognized that, I stopped feeling so angry. I could work with it. He sleeps much better in his own bed, so I made an effort to be home for his naps. Now I just expect that on a hectic day I will need to allow extra time to help him unwind at night. It does take effort, but when I do it, he sleeps, and when he sleeps, he's happy. If he's happy, the rest of the family is, too."

Know your child

The reality is there isn't one right strategy for getting children to sleep. You have to know and understand who has come to live with you. That's why it is so important to sit back and carefully observe what's typical for your child. This process allows your child to show you what he needs, when he's ready to try something new, and when he requires a little extra support. Your job is to watch, listen, and respond sensitively. Sometimes it necessitates waiting; frequently, it requires providing opportunities for practice; and occasionally, a gentle nudge to help him sleep well within the context of your family.

So, let's get a profile of your child's temperament. I'll show you what each temperament trait looks like and how it impacts sleep and effective strategies for managing it. You get to select those traits that are most significant, and the methods that fit best for your child and family.

GETTING A PICTURE OF YOUR CHILD'S TEMPERAMENT

Each of the five temperament traits can be placed on a continuum from mild to intense. Everyone has his own temperament, his own unique style. It's the overall picture that you need to take into account. There is not an ideal temperament, but positive and negative aspects of each.

As you review the traits, think about your child's most typical reaction. What responses have you come to predict?

Intensity

How strong are your child's reactions? Does his body tense quickly or remain calm? Does he cry loudly and vigorously or softly and mildly?

1 / 2 / 3 / 4 / 5

mild reaction	intense reaction
movements are smooth	movements are quick and jerky
squeaks when cries	wails
it's almost a surprise when he gets upset	a living staircase of emotion—up one minute, down the next
reactions are mild	every reaction is deep and powerful
usually works through a problem without becoming frustrated	easily frustrated
low-key	startles and gets keyed up

You know you have an intense child when you hear yourself describing her like Emma's mom did. "Everything about her is 'more.' She's the drama queen. If she's standing at the bus stop and her hat falls off, it's an ordeal. She's wired differently. Everything is higher, louder, and more intense."

And if you have an intense infant, he may demonstrate his intensity to you through his movements. That's what Samuel's dad saw: "He is constantly moving, sticking his feet straight out, and twisting his ankles around. He clinches up a lot more. His whole body gets very excited and he'll scream out. His brother never did that. The muscles on this kid are unbelievable. You can see them bulging in his thighs, because he's constantly tensing and flexing them."

If you selected a 4 or a 5, you can predict that your child will need more help from you winding down for the night.

How intensity plays havoc with sleep Children who are temperamentally more intense get upset faster and stay upset longer. Claire Novosad at Southern Connecticut State University has found that high intensity is related to sleep problems, either in getting to sleep or in staying asleep. These children are notorious catnappers, catching twenty- and thirty-minute naps, then awakening, even though they're still tired. They also tend to be alert more during the night and to call out to their parents, because they need help calming themselves and going back to sleep. Analysis shows that there is a higher level of the alerting transmitters in their system, which keeps them awake, while those transmitters that ease them into sleep remain lower.

Temperamentally intense children can sleep soundly as long as they are mentally and physically calm and their brain is set for sleep. Our task is to soothe them and gently teach them calming strategies to decrease their arousal level.

Effective strategies for managing intensity

Do not leave this child to cry

The advice to teach the intense child to calm himself by letting him cry does NOT work. Because of his physiological makeup, he has great difficulty calming himself, and can cry for hours, vomiting as his distress increases. This is not intentional or manipulative behavior on his part. He simply can't do it—yet. You can expect that this child, who needs to learn how to soothe himself and to fall asleep on his own, will require a much more extended process of breaking down the skills into tiny, manageable steps that don't overwhelm his ability to cope. He'll get there, but it won't be quickly, and it will take more effort on your part.

Provide touch

Intuitively, the intense infant quickly seems to realize that the adults in his life help him to calm his body. As a result, he seeks your company and doesn't want to be put down. It is this child that benefits most from the kangaroo- or soft-pack-style carriers, the daily massage, or the story read while he's sitting on your lap in a rocking chair. If the choice is his, he also prefers to sleep with you or on top of you, and, as he grows older, to have you sit near him or lie down with him. Your presence calms him. The key is to provide this kind of nurturing care but at the same time begin to introduce "soothers" that are not attached to you, and to practice with him moments (and I truly mean seconds) where you step away.

Jamie was a very intense baby. Initially, when his mom or dad would lay him down to sleep, he'd scream. "He would much rather have me there holding him, rubbing his belly and his legs to help him relax," his mother explained.

And so I suggested to her that she continue to stroke him, because that was what he needed, but at the same time to start introducing "sleep-time music" and giving him a little silky blanket to hold. Later, when we talked again, she said, "He's eleven-and-a-half months old now. He's still very intense, but now I can turn on the music, lay him down, and give him his blanket. He pulls it up over his face and falls asleep. Initially, that wasn't enough comfort for him, but now it is. Lately, he's been sleeping twelve hours a night without waking! It only took about a year. Of course, his brother never needed any of this—but that's the difference in temperament."

Protect the pace of his life

It's the intense child who really needs you to slow the pace of your life. She lets your family know by her inability to fall asleep that you are too busy. Her nap- and bedtime must be protected so that she doesn't become overtired. Once sleep-deprived, this child winds so tightly that she shrieks, the need for sleep fierce yet infuriatingly elusive. Being intense is stressful in itself, so, a day filled with a continuous barrage of stimulation and harried pace can be more than this child can endure. Her capacity to cope will increase as she grows older, but especially as an infant and young toddler, she needs to be sheltered a bit more. Rather than be a source of frustration, her needs can provide you with a great excuse to slow down your own life.

Allow time to unwind

"It takes a lot of time to help her wind down at night, and she can't unwind if she hasn't had any exercise." This is a very common report from parents, because time to unwind is essential for the intense child. Ironically, it may seem that while she needs it, she may resist it. If she's involved in a project, and her mind gets going, it can work on her body like caffeine. She pushes to continue, asking you to work with her, show her how something functions, or, if she's little, to simply pick her up and giggle with her. But what she really needs is enough time in the evening to allow her to talk through her concerns, write in her journal, get a massage, repeat calming prayers, or read, before she is expected to lie down and go to sleep. When she doesn't have the opportunity to unwind, you can predict that she'll awaken ninety minutes after falling asleep and be ready to go again, frustrated that you want her to sleep more.

Making calm energy a goal for your intense child pays big dividends, because the traits work together. Jim Cameron from Ounce of Prevention states, "As intensity comes down, your child also becomes LESS 'sensitive' and MORE adaptable." The tantrums disappear, cooperation increases, and, most important, your child sleeps soundly when together you manage the intensity.

Sensitivity

How aware is your child of slight noises, differences in tastes, textures, sights, or sounds? How aware is your child of the emotions of others? Does he react to certain foods, tags in his clothing, noises, lights, and/or your stress?

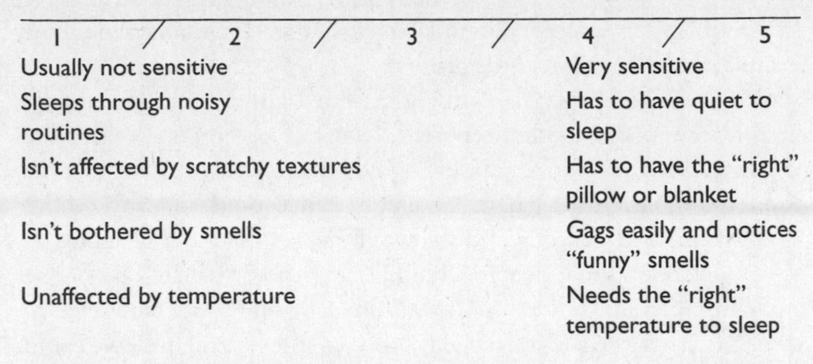

1	2	3	4	5
Usually not sensitive			Very sensitive	
Sleeps through noisy routines			Has to have quiet to sleep	
Isn't affected by scratchy textures			Has to have the "right" pillow or blanket	
Isn't bothered by smells			Gags easily and notices "funny" smells	
Unaffected by temperature			Needs the "right" temperature to sleep	

Place two four-month-old infants together in a room, then sound a loud bell. Odds are that one may not react at all or merely look in the direction of the sound. The other one, however, may startle, her eyes flying wide open, arms thrashing in the air before she bursts into tears. These are not learned responses. They are physiological responses. Some children, by their very nature, find it easier to block stimuli than others. That's why one child can fall asleep amid the noise of a birthday party, while another child awakens when someone merely tiptoes past her bedroom door. It takes much less stimulation for the highly sensitive child to feel agitated, increase her activity, or burst into tears. That's why she awakens when her diaper is wet, she's slightly hungry, or is feeling a bit lonely.

If your child falls into the 4 or 5 range on the sensitivity scale, you can predict that she will need you to protect her from overstimulation and to help her create a sleeping environment that feels right.

How sensitivity plays havoc with sleep

The highly sensitive child is keenly attuned to her sensory world and struggles to block out disturbing stimuli. As a result, she must sort through huge quantities of sensory information before she can feel safe and calm enough to sleep. She can hear the drone of traffic outside the window. The slightest ray of light can awaken her in the morning. The story of the princess who could feel the pea under twenty mattresses is not a fairy tale to a highly sensitive individual. And the breathing of her buddies at the slumber party really does keep her awake.

Fortunately, children who are sensitive but NOT intense seem able to shut down, and frequently will fall asleep in situations that are overstimulating to them. But those children who are intense AND sensitive find falling asleep a significant challenge, especially when the stimuli is intense or out of the ordinary.

Four-year-old Casey is a highly sensitive child. "She has always been like this," her mother reported. "She has never slept well. Little things bother her. She gets upset if the bath water is too warm or the towel too rough. If her pajamas aren't buttoned up right, or the collar is funny, she can't settle. When she was little, we used to leave notes on the front door so that people wouldn't ring the doorbell. The phone was constantly off the hook so it wouldn't ring and wake her. Once, at a holiday party, we had everyone go down to the basement so we could take her upstairs and get her down for a nap."

It's not just keen senses that may keep your child alert; she may also be highly sensitive to emotions. That was true for Emily. "She needs more cuddles and quiet whispering," her mother told me. "She has to say what's on her mind, and then she's done and can sleep. In the past, this was an issue for me. The moment when I was ready for her to go to bed, she wanted to talk about death, or life, and I would say we'll talk about that in the morning, and that bothered her and upset her."

Sensitive children are so aware that it's not just their own emotions that may alert them but yours as well. That's why when you take steps to reduce your own stress, your sensitive child benefits as well. But whether it's emotions or sensations that are keeping them awake, highly sensitive children need help blocking stimuli that alerts them.

Effective strategies for managing sensitivity

Believe them I had always assumed that my son had gotten his spirited tendencies from my husband's side of the family until one night I shared a hotel room with my mother. I watched as she very carefully pulled all of the blankets, except for the sheet, out from the foot board and then folded them two feet up from the end of the bed. Then she pulled a small pillow out of her suitcase—one I knew she'd had since I was a child—and tucked it under her chin as she crawled under the covers. When I asked her about her tactics, she said, "I can't sleep if my feet are too hot, and I need my pillow." When I asked why she had to bring her own pillow along, she patiently explained that the hotel's pillows did not "smell" right.

When your sensitive child declines the new nightgown you bought her because the lace "scratches" her neck, or complains that she can hear the television, even though her brother is listening to it with earphones—believe her. Truly, the world is a much richer source of stimuli for this individual. She is not trying to stall. She needs you to understand that her sleeping clothes have to feel right. And that she needs help blocking offensive sounds, smells, lights, and textures.

Take extra measures to reduce stimulation in the environment It is the highly sensitive child who is especially vulnerable to light and overstimulation. You can prevent this problem by closely monitoring the amount of time he is exposed to videos or television, even as background noise.

Recognize the importance of his "nest" This child, more than others, needs and likes his "nest" for sleeping. Leah was surprised when she began to notice the little things that could upset her son. "Whenever we visit my parents, my son sleeps poorly. This time, I realized it's the port-a-crib that is a problem. It's not even close to being the same as his crib. It makes different noises, and he doesn't like the mattress. This time I let him sleep with me. The mattress on my bed wasn't exactly the same as his at home, but there was no swishing sound, so he slept better."

The "nest" for the sensitive child has to include pleasant sensations AND block disruptive stimuli. What may surprise you is how negligible the offensive sensations may be that still upset your child. If there's a choice, give him the bedroom away from the street, on the quiet side of the house. Take special note of the weight, texture, color, and smell of his bedding. It's very important to him. Often he needs a heavy blanket, or just the opposite, no blanket at all. He needs to know that the doors and windows are locked. The room needs to be the right temperature. The light has to be just right, either completely dark or slight enough that he can discern the shape of objects yet not so bright that he is distracted by them. When it comes to night clothing, plan to cut out the tags, and check the waistband for the correct fit—that's if he will sleep in pajamas at all. Forget pajamas with the little feet in them. They're cute, but it's unlikely that he'll like them. He may prefer his favorite pair of socks and even a "night cap" that provides just the right input for his system.

You can make things even better by encircling him with body pillows, creating just the right pressure against his skin. It's the desire for "pressure" that often leads kids to want to climb in with you. When they are able to get between two parents, it's like two big body pillows, providing them with the sense of safety and comfort they need.

Consider working with an occupational therapist If, with all of these adaptations, your sensitive child still struggles to sleep, he may actually have a medical condition called sensory integration disorder. Consult an occupational therapist. She can design an individualized program of joint compressions, deep pressure, brushing, and other strategies that can help your child more effectively process the sensations coming into his body. These intervention strategies can be highly effective. After only three sessions, Molly's two-and-a-half-year-old son Tommy slept more than six uninterrupted hours for the first time in his life.

No child is going to need every adaptation. Experiment until you find the right sensory diet for your sensitive child to sleep best. And if you think this is all a bit unusual, check with a few highly sensitive adult friends. You'll discover that they have a very defined approach to "nesting" for the night, from a ritual checking of the doors to the turning off or on of lights, the massaging of a special lotion on their face and hands, or to the sensation of their partner's hairy chest against their cheek.

Those individuals who are less sensitive may find all of this information verging on ridiculous, especially if they can fall asleep on an airplane during takeoff, the couch in the middle of a family gathering, or, for that matter, any flat surface. Their ease in blocking sensations is an example of temperament, not willpower.

Adaptability

How quickly does your child adapt to changes in his schedule or routine? How does he cope with surprises?

1	2	3	4	5

ADAPTS QUICKLY	ADAPTS SLOWLY
Easily stops one activity and starts another	Cries or fusses when one activity ends and another begins
Is not upset with changes	May be very upset by surprises
Open to new activities	Is distressed by new activities or things
Usually complies with a request with little fuss	Immediately says "no" when you ask her to do something

There are significant differences in how individuals cope with change. Some children flow easily from one thing to another, while others struggle. You know you have a slow-to-adapt child if, when he begins to fall asleep in the car, you find yourself doing everything possible to keep him awake, knowing that you'll never be able to transfer him from the car to his bed. It's the slow-to-adapt child who insists on lying in bed for thirty minutes before he can possibly get up. Stopping the day's activities and beginning a bedtime routine can be a major upheaval. Returning to sleep after awakening in the middle of the night is difficult. And even though your child may have visited Grandpa and Grandma dozens of times, it's likely that he will still struggle to sleep well there, at least the first few nights.

That was the case for Brian. Ten-year-old Brian spent the weekend with his family at his grandparents' cabin. They arrived at one-thirty, unpacked the car, and immediately headed for the water. Brian and his eight-year-old brother and twelve-year-old sister played nonstop, jumping on and off the dock and running up and down the hill from the cabin to the lake. At five-thirty, they stopped for dinner, only to dash back to the lake for another half hour of swimming. The evening was spent in a competitive card game of thirty-one and making plans for the next day. At eight-thirty, seeing the trio drooping with fatigue, Brian's parents herded them to bed. There was no protest. They grabbed a quick snack, put on pajamas, brushed their teeth, and were in bed by nine o'clock. In fact, they were all in the same bedroom in the same bunk bed—two on the bottom double mattress and one on the single up top. At nine-fifteen, Brian's siblings were sound asleep. At one o'clock a.m., Brian was still wide awake.

Finally, with tears of frustration sliding down his cheeks, desperate for comfort and sleep, he crawled into bed with his mom and dad. Why did two sleep and one did not? Brian is temperamentally slower to adapt than his siblings. He needed more help shifting to sleep because of his physiology, not a desire for more attention.

How slow adaptability plays havoc with sleep Researchers have found that those individuals who are temperamentally slow to adapt have trouble shifting down, transitioning from one thing to another, and adjusting to new people and situations. Their slow adaptability is not limited to daytime hours. These individuals also have difficulty shifting from awake to sleep. If your child is temperamentally slow to adapt, you can predict that he will find it more difficult to cope with changes and transitions in his life. The key is preparing him.

Effective strategies for managing slow adaptability

Establish a routine

The switch to sleep for slow-to-adapt children is a major feat, so it's important to establish consistent bedtimes and awakening times so you set his body clock, making it easier for his brain to "switch" from alert to asleep. While this is true for all children, it's especially important for this child.

Prepare your child

It's your slow-to-adapt child who also needs clear cues and fair warning that the day's activities are winding down and it's time to move toward bed. By letting him know that he has ten more minutes until it's time to stop what he's doing, you can save yourself a hassle. Questions like "what do you need to finish in order to be ready for bed?" or "where would you like to save that?" help this child to let go of what he's doing and shift to the bedtime routine. You may also have to help him find a stopping point if his project is ongoing, with comments such as "three more turns" or "ten more pages." Dimming lights, picking up toys, pulling drapes, selecting books are all concrete cues that let this child know it's time to shift. Even with your preparation, your child may still react negatively. You can respond sensitively by acknowledging, "It's hard to stop and go to bed." Or, "You wish you didn't have to stop." But continue moving him toward bed. It's worth the effort, because as he gets more sleep, he will become more open to your direction.

Avoid rushing

Slow-to-adapt children do not like to be rushed or surprised. In the morning, if it takes your child thirty minutes to come to full arousal, allow that time in your routine instead of valiantly trying to change his ways. The slow-to-adapt child is not going to jump out of bed. He needs to lie there for a while and slowly "switch" to awake. You'll both be much happier if you recognize it, expect it, and plan accordingly.

It's easy to allow the slow-to-adapt child to sleep in until the last minute, because it's such a struggle to get him going, but when you wait until the last minute, you end up feeling rushed. Rushing increases intensity. As intensity goes up, adaptability goes down. Instead of getting him going as you had hoped, it actually slows the process down as he "shuts down," overwhelmed by the tasks that face him. A few extra minutes in the morning can mean the difference between starting your day out in calm, rather than tense, energy.

Avoid skipping or changing the routine

Slow-to-adapt children also like predictability. That's why if, at bedtime, you are running a few minutes late, and wish to shorten his normal routine by dropping one book or skipping snack, it's likely that this child will be upset by the change. It is very important to establish a clearly defined routine that is simple and easy to stick to, even on the hectic days.

If it is necessary to shorten the routine, select a book with fewer pages, or prepare a simpler snack. In that way, you avoid trying to skip a segment of the routine, which is the real trigger for the slow-to-adapt child.

Make it a priority to be home for bedtime and naps

Slow-to-adapt children just do not sleep well in new or different situations. While a quick-to-adapt child may be easily shifted from the couch to his bed, or fall asleep on the bed or under the table at Grandma's house, that's not going to happen with a slow-to-adapt child. He's not being stubborn, he just can't do it. By making it a priority to be home for naps and bedtime whenever possible, you ensure that your child will get the sleep he needs. While this may seem to be a frustrating restriction, the reality is that your slow-to-adapt child will be much more flexible, if he gets his sleep.

Regularity or predictability of body rhythms

Is your child quite regular about eating times, sleeping times, amount of sleep needed, and other bodily functions?

1 / 2 / 3 / 4 / 5

REGULAR	IRREGULAR
Falls asleep at the same time almost every day	Never falls asleep at the same time
Is hungry at regular intervals	Is hungry at different times of the day
Naps for the same amount of time each day	Naps vary widely in length

I'm choosing to alter my pattern by addressing the needs of both regular and irregular children, because this trait is so important when it comes to setting the body clock.

When your child's body rhythms are very regular It's the regular child who asks to be put to bed, or falls asleep at the same time every night, no matter where he is. You never have to put him on a schedule, because he does it himself. Regularity can override high sensitivity and intensity. No matter how stimulated or excited this child is, when his clock says time for sleep—he sleeps. It can be a joy, but the challenge for the regular child is trying to delay bedtime or "switch" his schedule. Any change poses a major problem.

Effective strategies for managing regularity

Honor your regular child's body clock

What the regular child needs you to know is that he really does need to sleep when his body says "sleep." He's not being inflexible to irritate you; it's simply that his clock is so strong. When his bedtime is delayed, it's unlikely that he'll be able to sleep in the next day, because his clock will awaken him. As a result, he becomes sleep-deprived when his schedule is disrupted. Because his clock is so defined, it also can take him up to three weeks to recover after springing forward for daylight savings time, and he is more affected by jet lag.

When your child's body rhythms are irregular If you find yourself staring aghast when a child says, "Mommy, I want to go to bed," because your child would NEVER say those words, it's likely that you have an irregular child. The irregular child is unpredictable. You don't know when he'll need to nap or want to fall asleep. While the flexibility can be delightful, the challenge is that this child can easily become sleep-deprived.

How irregularity plays havoc with sleep The temperamentally irregular child appears to have difficulty setting his clock. As a result, his brain never quite knows when to "switch" into sleep, unless the adults in his life help him to create an environment that cues him.

Effective strategies for managing irregularity

Gently nudge your child into a schedule

Sensitively responding to your child is essential for connecting with him, but when your child is irregular, "following his lead" completely can lead to a schedule so chaotic that no one gets any sleep. I am NOT suggesting a rigid feeding schedule, nor would I ever suggest you let your child cry himself to sleep. Instead, the plan is for creating a structure that gently moves him toward a schedule that fits him AND your family.

Ryan was seventeen months old when his mother called me looking for help. His tantrums were so intense that his parents were beginning to avoid public situations with him. After reviewing with her a typical schedule for Ryan, I realized that he was severely sleep-deprived. "I've been lying to my pediatrician since he was born," she admitted, "because he's never been on a schedule. But nothing has seemed to work." I encouraged her to begin by establishing a regular wake time for him, and took her through the process of setting his body clock, as I showed you in chapter 9. Three weeks later, we talked again. "I never forced him," she said. "That doesn't work with Ryan. But I did establish more of a routine. If he fought it, I calmed him but I wasn't swayed. If he needed comfort, I gave it to him. If he needed to be near me, I stayed with him, but I insisted that he wasn't leaving the bedroom or watching a movie, it was time for sleep. Instead of sleeping nine or ten hours, he's now sleeping thirteen. The change in his behavior is almost miraculous, and not only is he sleeping better, he's eating better as well."

It's somewhat ironic that the child who needs the schedule the most strongly resists it. Your irregular child needs the environment around him to help him set his body clock. Without that assistance, his system can run wild. By establishing a routine, you ensure that he gets food when he needs it, takes time to unwind, and, most important, gets adequate sleep. These changes help him manage the tension and, as a result, prevent the tantrums, because he has the energy to cope. Without the supportive structure, the irregular child stays up, which creates tension. As tension increases, he seems to try to escape by watching television, which keeps him up later. He also tends to seek carbohydrates and caffeine to keep himself going, thus creating the cycle that sets him up for sleep deprivation.

You cannot, nor would you want to, force your child into a schedule. Instead, it's a matter of creating the routine and providing the support to gently move him into it. The second challenge is that once you have him on a schedule, you have to keep to it seven days a week, because it is this child who will "fall" off immediately, because his temperament isn't helping him stay in sync. But when he does become ill, or if you travel, or something else disrupts your schedule, at least you know now what you're dealing with and can take the steps to bring his body back into a rhythm.

Energy level

Is your child always on the move and busy or quiet and inactive? Does he need to run, jump, and use his whole body in order to feel good, or is he happy playing quietly?

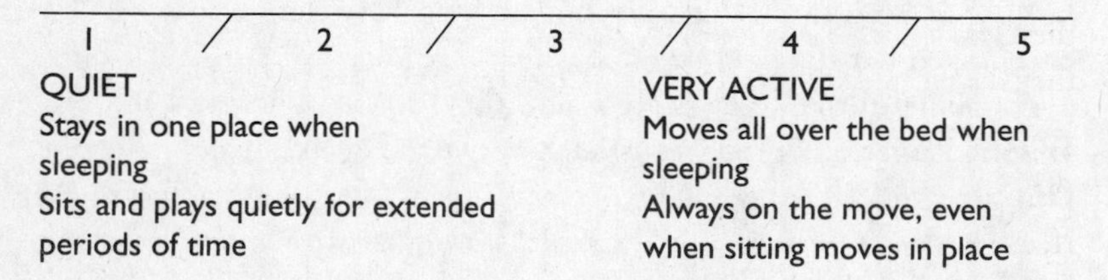

It's the high-energy child who hates to stop and go to bed. "Tatum has no cues that she's ready for sleep," her dad told me. "She's like a windup airplane, flying around the house then suddenly collapsing. She can be talking and fall asleep midsentence. She gives it her all and then there's nothing left."

How high energy plays havoc with sleep The high-energy child is notorious for her short "window" for sleep. Because she is so active, it's easy to miss the cues that she's growing fatigued. There may be only a very slight slowing of movement, or droop in her posture. If you miss it, her system will charge up again, and she will be alert for another ninety minutes. She's so wired that it's hard to believe she's actually overtired. It's also often difficult for her to read her own cues. She doesn't want to miss out on anything, and so she pushes to keep going until she's overtired and too tense to sleep. An unfaltering routine not only helps to earmark her "window," but also assists her in setting limits for what she can accomplish on a particular day.

Strategies for managing high energy

Provide exercise

Everyone benefits from exercise, but it's the high-energy child who needs it the most. "On the days that I've been working at home and Nate has just been playing quietly in the room with me, he doesn't sleep well, and it takes him much longer to fall asleep," his mother told me. Physical exercise during the day is a must for high-energy kids; otherwise they struggle to unwind at night.

Mental exercise matters, too

Sometimes it's not just physical exercise but mental exercise that the high-energy child needs in order to be fatigued enough for sleep at night. Stephanie Tolan writes, "When my sons were challenged in school, they stayed busy until eight or nine at night, but when they were not challenged, they insisted on staying up until ten or eleven. It was as though they needed a certain amount of intellectual or creative exercise, and if they didn't get it during the day, they pursued it well beyond bedtime."

When your high-energy child won't stay in bed, consider visiting your doctor

If, as your high-energy child becomes drowsy, he can't stop moving his feet, or complains of unpleasant sensations in his legs, such as creeping, crawling, itching, or burning, he may have a medical condition called Restless Leg Syndrome. This condition often results in difficulty falling or staying asleep. The sensations usually occur in the calf area of the leg, but may be felt anywhere from the thigh to the ankle, and occasionally even in the arms. The sensations also often begin whenever the individual sits down at a desk, rides in a car, or watches a movie. Symptoms tend to be especially prevalent at night and may occur on a regular basis or only intermittently.

Due to the discomfort, a child with RLS may feel an almost irresistible urge to get up and move, and, as a result, refuse to stay in bed. Walking, rubbing the legs, or massage helps to relieve the symptoms.

Many individuals with restless leg syndrome also have a sleep disorder called periodic limb movement (PLMS). This condition, according to information from the online medical library, is characterized by involuntary jerking or bending leg movements during sleep, which occur every ten to sixty seconds. These movements can prevent your child from experiencing sound sleep.

If your child complains of these symptoms, see your doctor. If he or she is not able to help you, consider a visit to a regional sleep center. According to the National Sleep Foundation, restless leg syndrome and periodic limb movement are very frequently under-diagnosed in the primary-care setting.

Review your child's temperament picture Now go back through each of the temperament traits and total your responses. Mark your total on the scale below.

SCORE:

5–10 EASY SLEEPER 11–17 MODERATE SLEEPER 18–25 SENSITIVE SLEEPER

This is not a scientific analysis, it is merely a tool to help you get a picture of your child's temperament and the way it impacts his ability to sleep. You can use it as a guide, helping you to understand your child and the challenges he faces to sleep well. If, for example, your child is slow to adapt, you can predict that moving him from a crib to a "big-boy bed" may require practice before he's expected to sleep there. Instead of getting frustrated with him, you can give yourself permission to work with him. You will know that you are not doing something wrong but rather that he simply must work harder.

Temperament is a tool that allows you to recognize what your child is experiencing and, as a result, select the strategies that FIT him. By doing so, you send a message of acceptance and love, while at the same time teaching him the skills to successfully get the sleep he needs. Understanding temperament allows you to love your child for who he is and affirms your sense that this child needs more to sleep.

(If you'd like to know more about temperament, see my books *Raising Your Spirited Child* and *Raising Your Spirited Child Workbook*.)

ENLISTING EFFECTIVE STRATEGIES FOR SOUND SLEEP AND GOOD BEHAVIOR

ELEVEN

ENDING THE BEDTIME BATTLES

"The two most important things parents can say to their kids are, one, 'I love you' and two, 'Go to bed.' "

—Charles Osgood

I must be honest. When my children were young, bedtime was frequently a frustrating event in our family. I have always worked two nights a week, which meant that when my children were young, their dad often put them to bed. When he did, the kids got a Chief Tony Bouza story. Tony was the Minneapolis chief of police at that time, and my husband would regale the kids with tales in which Tony called upon them to help solve a particularly puzzling crime. Josh thrived on the tales. Kristina, while intrigued and delighted by them, too, implored, "Don't put me in the story. Don't put me in the story!" (We laugh about that today, since Kristina is now our mountain climber and sky diver.) Frequently, when I arrived home at nine-thirty p.m., I was not pleased to discover two very excited children still wide awake. My husband was not intentionally plotting to irritate me, even though it sometimes felt like it. Instead, he was innocently enjoying the time with his children. He did not realize that Tony Bouza stories and wrestling matches wreaked havoc on bedtime.

I wish I could say that when I was home, everything went smoothly, but that is not true, either. If they needed it, or there was time, I'd begin the bedtime routine by bathing the kids, but not always. I didn't know about the importance of a predictable "transition" activity, and so, there was often a tussle to get them moving. Our schedule was also erratic. The kids were involved in sports, which meant that many evenings were spent at practices and games. Sometimes we put them to bed at eight o'clock, and, then again, it might be nine o'clock or nine-thirty. We were clueless to the fact that our inconsistent bedtimes were disrupting their body clocks.

Fortunately, Kristina, who is temperamentally very regular, weathered our lack of consistency relatively well. Her body clock was strong enough that if we were driving in the car at eight o'clock, she simply went to sleep and could be dropped into her bed when we arrived home. Josh, however, didn't have that same strong body clock, and struggled to fall asleep. We wondered why, and, quite honestly, were perturbed by how much more effort it took to get Joshua to sleep. I spent years sitting by his bedside waiting for him to fall asleep, wondering if I'd still be doing it when he was an adolescent. Ironically, at seventeen, Josh would frequently ask me, "Mom, aren't you going to bring me a glass of water and talk to me?" I would, and we would chat about the events of his days as he settled in for the night. Silently, I thanked my lucky stars that my adolescent son was talking to me about his life. The payoff of all those hours was mine!

I share all of this with you because, in hindsight, and with a great deal of research and practice with the families I work with, I've discovered two very important things. While we cannot make our children sleep, we can ASSIST them to sleep with an effective bedtime routine, and that bedtime, rather than an event to dread, can be a time to enjoy—an opportunity to connect with your child and build a relationship that keeps you working together for a lifetime. It begins by recognizing that children are more like jumbo jets than helicopters.

Preparation and a gradual descent to sleep make a difference

Landing a jumbo jet is not a simple process. Miles from their destination, the pilots begin to prepare. They check the weather, determine which runway to utilize, the level of instrumentation to use on approach, as well as the optimal speed. Once those decisions are made, they start to configure the aircraft appropriately, and inform the flight attendants of their progress. They even come up with a contingency plan in case there is a missed approach.

Back in the cabin, the flight attendants prepare the passengers by announcing, "Ladies and gentlemen, we are beginning our descent. At this time, please turn off all electronic devices and store all items in the overhead bins, or under the seat in front of you. We expect to arrive at our destination in twenty-five minutes."

What the crew is trained to know is that conscientious preparation and a gradual descent lead to a soft landing and satisfied customers.

When it comes to bedtime, most children are like those jumbo jets. Their days are often spent "flying" from one activity to another, and they need to gradually "glide" from the "high" of their day to a "soft landing" in bed. It's true, there are a few "helicopter" kids, like Kristina, the ones who can be plucked from a party or a soccer practice and dropped into bed, where they promptly fall asleep, but they are rare. Most children need a little more preparation. When they get it, they can easily slip into sleep, and everyone is happier. Initially, we are the "crew." Ultimately our children will take over the task themselves, but until they do, we get to configure the plan.

A positive routine calms the body for sleep

In a study completed at the Arkansas Children's Hospital, researchers found that parents who engaged in four to seven enjoyable activities, lasting no longer than twenty minutes, BEFORE expecting their children to fall asleep, eliminated the bedtime battles AND reported improved marital satisfaction. The positive strategies were even more effective than leaving the children alone to cry themselves to sleep.

"That's so true," Beth exclaimed. "Last night, the girls had dance practice, so we were running late. I tried to hurry them into bed by skipping showers and reading. But two seconds after lying down, Sasha, the seven-year-old, was sitting back up. 'Mom, I can't sleep. Could we read, just a little?'

"I had other things I needed to get done, and I didn't really want to spend another twenty minutes sitting there. But I realized she needed something, so I told her, 'We can read five pages,' knowing she'd want ten. We ended up agreeing on seven, which is what I'd expected. She lay quietly and listened. When I finished, she asked me to turn on her night-light as I left, which I did. I know it's what she needed, but I always feel a little guilty that somehow I'm letting her manipulate me or I'm not teaching her to be independent."

Beth had discovered the importance of helping her daughter wind down before sleep, and found a way to do it that respected both her daughter's needs and her own. By taking those few minutes, she helped her daughter to relax enough to be able to fall asleep on her own. Yet she worried that she was doing something wrong.

There's a great deal of emphasis on teaching children to "self-soothe" to sleep, and research demonstrates that children who pass independently through the "window" to sleep, sleep more continuously through the night, but "self-soothing" does not mean dropping into bed without any preparation.

By recognizing the need for a gradual descent from alert to asleep, and planning for it, you position your child to catch his "window" for sleep. As a result, he is ready to pass through that window on his own, his body relaxed and his brain cued that it is time for sleep. Without that preparation, his body may be in a state of tense tired, too wired to "switch" to sleep. That's when the battles begin, and why it is so essential to recognize that what happens BEFORE your child lies down to sleep is just as important as putting him to bed.

Every child's transition to sleep is slightly different. But if you listen and watch carefully, your child will show you what she needs, and when she's ready to "switch" into sleep. Sensitively responding to her does not mean that there are no limits. There are, but you are working together with your child, helping her to feel calm and safe enough to sleep.

I will provide you with a guide for an effective bedtime routine, but just like the pilots make adjustments for the weight of their plane and the weather they're flying in, you will have to adjust the plan to fit your child and the conditions you are working within. There really is not one right way for a child to arrive in the land of sleep. There are different approaches; the key is to find the one that fits your family. It requires a bit of creativity on your part, a willingness to recognize the individuality of your child, and occasionally going against the crowd.

AN EFFECTIVE BEDTIME PLAN

Going to bed is something you do every single night. It's a repetitive process that allows you to predict how everyone is going to respond, because you've been through it before. If you can predict it, you can plan for success. You can use your knowledge of *tension, time,* and *temperament,* the words and the strategies you've learned to ensure that the day ends with everyone feeling safe, tranquil, and ready for sleep. I've designed three simple steps that you can follow. They are:

1. Seeing potential trouble spots
2. Establishing a routine that calms and cues your child's brain for sleep
3. Tweaking the plan for individual needs and potential trouble spots

Followed closely, these three steps will allow your child to be SET for sleep.

Seeing potential trouble spots

Seeing potential trouble spots means approaching bedtime with one goal in mind: a goal to connect, calm, and cue your child for sleep. Instead of approaching bedtime with trepidation, you can expend your energy planning for success by preparing your child's body and her brain. If you're groaning at the mere thought of "planning for success," knowing that by that time of the day, you are much too drained to even think, you may be delighted to discover the process begins with taking care of yourself.

Check your own level of tension and fatigue Sleep comes most readily if no one in the family is tense, angry, or fearful. Just as you get to set the tone for how the day starts, you also can influence how it ends. Everyone has a "grumpy" time of the day. If you can predict that your patience will be frayed and that you will be silently praying, "Please, just for once, let them fall asleep quickly," as you approach bedtime with your children, it's essential to take a few minutes for a mom or dad "time out." By doing so, you can move yourself out of tense tired into calm tired. Otherwise, your children will catch your tension, and the night will unravel from there. But how do you grab a few moments for you? That's what I asked the parents in my class, and here's what they had to say.

"For me, it's a dinner where we actually sit down and talk. We're not rushed. And even after the kids leave the table, Deanne and I sit for a few more minutes, drinking coffee or tea. I think it's good for the kids to see us together, and they often come back to sit with us."

"I take a shower while my husband plays with the kids. Sometimes I stay there for a very long time!"

"I used to go nuts trying to get the boys to brush their teeth. They're old enough to do it on their own, so, now I just send them upstairs while I sit downstairs and read for ten minutes. I'm much more patient if I take a moment for myself, and then we don't end the day yelling at one another."

"When it was time to put the kids to bed, I'd think to myself, it will take ten minutes to read to them and ten minutes to get on their pajamas. But then they wouldn't cooperate, or David would be banging Lizzie on the head. I'd get upset because they weren't moving fast enough. Now my goal is for bedtime to be a good experience. In order for that to happen, I can't feel rushed. Now we go upstairs an hour before lights off. That way I'm not sending them vibes to hurry up and go to sleep."

"I put three kids to bed on my own. There are very few opportunities for me to take a break, so, the first thing I do after the older ones have completed their homework, is sit down and read aloud to them. Snuggling and reading relaxes me as much as it does them."

Whether it's a phone call to a friend or favorite relative, a few minutes talking with your partner, a break without anyone else bothering you, a few deep breaths, snuggles, or allowing extra time, calming yourself makes everything go more smoothly. So, tonight, stop and check your tension level. Are you bubbling inside? If the answer is yes, give yourself permission to quit what you are doing and instead take the steps to compose yourself. Better yet, make step one in your bedtime routine fifteen minutes of downtime for you. Then you'll be ready to help your children prepare for sleep.

Check your child's level of tension and fatigue Review again the signs that your children are missing sleep. Then select a specific time of the day, like right after school, when you pick them up from child care, or at dinner, to mentally check the tension and fatigue level. Are the kids irritable and whiny? Is a homework assignment overwhelming them? Are they clumsier than usual, or begging for treats? How is their focus? Are they wandering from one toy to another, not listening or needing more help and attention than usual? Can they get along with one another or are they at each other's throats?

If your children are in a state of calm energy or calm tired, you will NOT see the "misbehaviors" tied to tense energy and fatigue. Even though it's near the end of the day, they will still be playing well with one another, listening to you, cooperating, and being flexible. You can feel comfortable moving ahead on your normal schedule and with your evening plans.

But if your child is struggling to manage his emotions, body, focus, or to work with others, it's critical that you recognize that he is in a state of tense tired. The evening ahead is likely to be rough, unless you take steps to change the path.

Derrick reflected on his experience. "Last night, when Leanne came to the dinner table, the power struggles started," he told me. "She didn't want to sit in her chair. She didn't want to eat what the rest of us were eating. Trying to distract her, I asked her if she wanted a bubble bath tonight. Even though she loves her bath, she told me she hated it and didn't want to get her hair wet. I felt myself getting angry, and then, all of a sudden, I realized, she's not making any sense. Nothing pleases her. This is what happens when she's overtired."

In case you are not quite sure that these behaviors are being fueled by tension and fatigue, review your day, paying special attention to those things that create tension or disrupt the body clock. Ask yourself, what's been happening? Has the pace of the day been hectic or the stimulation level high? Was there a different teacher today? Are new developmental skills evolving? Have there been more separations than usual? Is an exciting or frightening event about to occur? Has sleep been disrupted, a nap skipped, bedtime delayed, or an early awakening occurred? Did exercise time get skipped today? Whether positive or negative, these events create tension and upset the body clock, impacting what happens when it's time to fall asleep.

Work together When you are aware that your child is tense and out of sync with his body clock, and understand the reasons for it, you won't feel as though your child is intentionally misbehaving. You can forecast ahead and choose to adjust plans accordingly. If your child is an infant, you'll be making the decisions about altering plans. But if your child is older, you will be helping him to become aware of what's happening inside of his body by talking about it. You can help him see potential trouble spots by sharing your observations and asking questions, such as "I see you struggling to stay focused on your homework, would it be helpful to go to bed earlier and finish it in the morning?" Or, "You've got a soccer practice that will run late tonight. What could you do before you go to make it easier to fall asleep afterwards?"

Then, together, you will decide what to do. You may choose to leave practice early, take time to select a short book and place it ready at his bedside, read and relax before practice, choose not to attend the practice and instead start the bedtime routine earlier, or plan to skip bath and have more time to unwind by reading or massaging. By predicting and planning ahead, you'll avoid being caught off guard at bedtime, frustrated by a child who is too tense or out of sync to slip into sleep. Jane found a way to do this.

"Anna jumped into the car after school, chattering nonstop. Damon immediately shouted, 'Shut up!' And then the two of them went at it. In our family, you can say 'please be quiet,' but 'shut up' is not acceptable. I knew right then it was going to be a bad night. I'd planned on stopping to pick up a birthday gift, but I recognized it wasn't going to work. Part of me was ticked off. I get tired of their fighting and wanted to just scream, 'STOP!' But somehow I had the sense to take a deep breath and ask them, 'Do you guys think you can make it through the store or are you too bubbly inside?' 'Can't we do it tomorrow?' Damon pleaded. I was in a quandary. If I didn't go to the store, it would feel like I was giving in, but then I remembered that just the day before they'd both been very cooperative when Anna's dance rehearsal had run late.

"I took them home and gave them a snack. Damon went off to play quietly in his room. I was still frustrated by the change in plans, and Anna was driving me wild, following me everywhere and talking incessantly. I walked past the treadmill and realized that I'd skipped exercise, which wasn't helping my tension level any. So I told her, 'You need to talk, Damon needs time in his room, and I need to exercise. Get your markers, pens, and paper. You can draw and talk while I walk on the treadmill.

"Forty-five minutes later, she had finally wound down, and I was feeling more energized. I made dinner, then, thirty minutes early, took both of them upstairs to get on their pajamas and to read to them. When Damon asked for a massage, I was prepared. If I hadn't allowed us that extra time, I would have been irritated and tried to talk him out of it, which would have meant a battle. But I'd expected it, and, after only a few minutes, he was ready for sleep.

"Meanwhile Anna was willing to read on her own, because we'd had time together. When I went back to her, she turned off the light and lay down without a fuss. If I hadn't let her talk earlier, she would have fought to keep me there.

"The next morning, I reminded them that we would be picking up the gift after school. We did, without any hassle. In the end, it all worked out. I just have to remind myself that it's all right to listen to my kids and to change our plans to fit their needs."

Jane recognized the signs of sleep deprivation and understood the needs of her children as well as her own. As a result, she was able to predict the need for an earlier bedtime and plan time for additional soothing, calming activities. By doing so, she was able to prevent not only a bedtime battle but a potential argument at the shopping center as well.

In the future, our goal would be to teach Anna and Damon to note their own fatigue and tension and to know what they need. We'll teach them phrases like "Dad, I'm really tired, could we please just go home." Or, "When I'm tired, I need quiet." Or, "Mom, I think I'm going to need help winding down tonight." (They really DO learn to say these things.) When they recognize their own fatigue behaviors, they can respectfully and appropriately get their needs met. Bedtime battles disappear, because you move into the routine with everyone feeling calmer and more relaxed, even if the day has been a tough one.

You can start a successful bedtime routine by respecting the need for sleep, thinking about it, and planning ahead. Continue it by establishing a predictable routine that cues your child for sleep.

Establishing a predictable routine

I asked six-year-old Tommy and four-year-old Michael, "What's the first thing you do when it's time to go to bed?"

"Have a snack," Michael responded quickly.

"Then what?"

"Put on pajamas and brush teeth," Tommy offered.

"Then what?"

"Read a story," they added in unison.

"And?"

Quickly, they recited, "Go to bed, get the covers up, get a kiss, turn off the lights, and then we have a hug and then we have a night-light on, and then Mom says, 'Go to sleep!' "

An effective bedtime routine is so simple and predictable that even young children can describe it. It includes steps that you can follow consistently for years, no matter where you are, or whether it's an easy night or a crazy one. It includes activities that all the adults caring for your child can feel comfortable carrying out and that eventually can be taken over by your child herself. Most important, it leaves your child feeling calm and secure, ready to slip into sleep—ultimately, on her own, in whatever space your family chooses.

Even when your child is a young infant, you can begin to create a routine that calms and cues him. That's not to say he'll follow it yet. We'll be working together with him, learning to recognize what he needs, while, at the same time, beginning the gentle nudge that helps him to discover how to sleep in your family.

I've developed a four-step routine that I use when working with families, and I've created a little slogan to help you remember it. It is: Tense Children Can't Sleep. Use the first letter of each word to remind you of the steps.

1. **Transition Activity:** The transition activity marks the beginning of your bedtime routine, such as snack, dimming lights, or picking up. It is an action or announcement that you can initiate at approximately the same time and repeat every single night. Its occurrence clearly indicates that active time is finished and it's time to begin preparing for sleep. Check the "Goal/Sleep/Activity chart" you created in chapter 9. The time you selected to begin your bedtime routine is the moment to announce your transition activity.

2. **Connecting and calming activities:** Connecting and calming activities soothe your child so that he is relaxed enough for sleep. They're varied. Some are done together with you and include things you enjoy as much as your child does. Others are carried out independently by your child.

3. **Cue Activity:** The cue activity is also consistent every single night. It's the song, the kiss, the prayer, the turning on of a night-light or a fan, which signals it is now time to go to sleep.

4. **Switch Activities:** Switch activities are the last memories and

sensations experienced as your child falls asleep. They mark the sleep time on your ideal sleep schedule.

Let's take a closer look at what each step looks like.

Transition activity Anyone who has ever flown in an airplane knows the illumination of the seat-belt sign and the announcement to stow away tray tables and put seats in an upright position signifies that the plane is about to land.

Your children, even infants, need a similar predictable proclamation or event that occurs at the same designated time (or very close to it) every single night. It can be an announcement to clean up, a song or a bath, bedtime snack, story time, putting on pajamas, dimming lights, turning off all electronics, heading to the bathroom for toileting and teeth-brushing, or going upstairs. What's essential is that it is a clear event that signals to your children that it's time to shut down, stop the activities of the day, and begin preparing for the approach to the land of sleep. Look back at your goal sleep chart for the designated bedtime you chose. It may help you to be more consistent if you make it a time where there is a natural transition, such as the ending point of another evening activity. That's what Wendy did.

"While Hannah finishes her homework, Micah, my three-year-old, will play quietly in his room. The baby sits in his high chair and eats a few Cheerios while I help Hannah. By seven-thirty, Hannah has completed her assignments. I get the baby into his pajamas and ask Micah and Hannah to change into theirs. They don't fight it. I guess that's because we always do the same thing every night, and they know what's coming next—story time—which they love."

Think about the number of children that you have, their ages and their interests, and whether or not they're all going to be on a similar schedule. Then choose your "transition" activity.

If it's difficult for one or more of your children to stop what she's doing, select a fun activity like story time, rather than toileting and teeth. The story will calm her and make it easier to move on to the more challenging tasks. If a bath soothes and calms, it can be your "announcing" activity, but if you tend to bathe your child inconsistently, or if it revs rather than calms her, don't use it. Instead, choose another time during the day for a bath. Or, if your child becomes drowsy while you read, you may choose to complete the tasks of pajamas and teeth first, so that you don't have to disturb her. What's most important is that the activity works for your family, allowing you to begin the routine with a clear signal and without an argument.

Once the transition activity has unmistakably begun the shut-down of the environment and eased your children into preparing for sleep, you are ready to move on to those activities that connect and calm.

Connect and Calm Activities "Don't leave me." "Sit with me." "Talk to me." "Scratch my back." "Can't we read one more story?" Children want time with you, and if every night you find yourself fighting because your child doesn't want you to leave him, expect it and plan for success. When you include a time to connect, you'll discover that it's much easier for your child to let go and fall asleep. The good effects will spill over into the next day not only in the improved mood but in your child's willingness to work with you. These benefits are true not only if your child is young, but for adolescents as well. Studies show that teens whose parents continue to monitor bedtime get more sleep than those whose parents don't. And those same teens are more willing to work with their parents, because of the strong relationship those nighttime conversations foster.

A few minutes connecting and calming can make a significant difference in how difficult it is for your child to fall asleep. When I checked my e-mail, I was delighted to find a note from Sylvia. "Tonight at bedtime I promised myself that I was only going to treat Steven in a way that 'connected and calmed' us. I have to tell you, I think a miracle just occurred. I am writing this at nine o'clock p.m., and Steven, who is three-and-a-half and a notorious night owl (usually ten-thirty or eleven o'clock p.m. bedtime), is asleep in his room with his door open and his night-light on. He allowed me to leave the room, played with his bear, and looked at his books for twenty minutes or so, and then wound down and fell asleep."

Each child is unique Everyone is different. We don't all find the same activities soothing and calming. Some of us like touch, others prefer space. Some like to read, others would rather tell stories or talk a bit. There isn't one right calming activity for you and your child. It's a matter of working together. Select activities that you look forward to as much as your kids do.

Jason was laughing when he told me, "My wife and I had to select different children's novels to read to the kids, because neither one of us wanted to miss any of the story. When I put the kids to bed, I read mine. When she does it, they read hers. I actually look forward to my turn."

For Jacob, it's the stories he tells the kids. "We all sit on Isaac's bed and I tell them a story. It's usually just a story about their day, but they love it. Like, 'Once upon a time, there was a little girl named Sarah, whose best friend was moving away. Sarah was very sad.' It provides a great opportunity to work through things with them."

You'll also want to include soothing activities that don't involve you, like looking at books, writing in a journal, playing with "quiet" toys, drawing, listening to music, watching fish in an aquarium, or snuggling a favorite stuffed animal or blanket. Not only does this allow your child to learn relaxation skills, but it also provides you with more flexibility. If you are putting more than one child to bed, the one you are not with still has a soothing activity to do while waiting for his turn with you. Jenna found this particularly helpful.

"My four-year-old son was thrilled with the idea of looking at books in his bed. I was amazed that he could stay on task while I put his sisters down. He knew his turn was coming and was willing to wait. The very best part of the process was that I made a connection with each child before sleep, which hasn't been the case for months."

"Lovies," like a silky blanket, soft teddy bear, or fluffy pillow are also great "self-calming" items. I like objects like these, because of their portability, and because they can be continued to be used by your child as he grows. You'd be amazed to see the number of special blankets, pillows, and even stuffed animals that arrive at a college dormitory. Even guys have "favorite pillows"!

Introducing "lovies" Studies show that 80 percent of young infants also use some type of "lovie." However, the "lovie" tends to be inconsistent and to change over time, so, you may not even be aware of its use. Young infants, three months and younger, tend to use their own hands. While you might worry that your child is a thumb sucker, there is a benefit. Thumb suckers are reported to sleep better. Six-month-olds tend to prefer a cloth that has their mother's odor on it. A nightgown or a T-shirt you've worn may be perfect.

You can introduce a "lovie" to your baby by draping that silk blanket over your shoulder while you feed or burp her. Not only will she begin to associate that blanket with the warmth of you, it will also hold your smell, which soothes and calms her. Even if initially she doesn't seem interested in it, continue to use it. She may surprise you in the future.

Changing the "lovie" when the "lovie" is you Infants may also choose you for their "lovie." If you have long hair, make a decision. If it hangs down while you're feeding him, it's easy for your baby to reach up and stroke it. There's nothing like mom's silky hair to soothe and calm. But if you don't want to be your child's "lovie," you may want to pull your hair up. If, however, you love that sensation of nursing your baby while she strokes your hair, feel free to do it, knowing that one day, like Stacey, you may be ready to change it, and you can.

"When Emma turned eighteen months, there was more hair pulling than stroking going on, and I was ready to be done. But she was so used to falling asleep with my hair in her hand that she struggled without it. So I cut my hair, banded it together, and gave her a portable Mom's ponytail. It lasted about three weeks, just long enough for her to make a transition to a teddy bear that was almost as soft."

When you wean your child, replace suckling with soothing and calming For many children, weaning means the loss of their best "connect and calm" activity. Even if prior to weaning, your child hasn't gone to sleep "sucking," she has had the opportunity during the day to manage tension by lying back, relaxing, and sucking. Repetitive motion, especially repetitive motion of the jaw, produces stress-relieving hormones. Once the bottle, breast, or pacifier is no longer part of your child's life, a very effective tension-buster becomes unavailable to her. So, when you wean, don't stop snuggling and holding your child. Offer her straws when drinking from a glass, water bottles that require "sucking," and chewy foods, to foster the oral repetition that she has been accustomed to in the past. Expect to extend or add another soothing activity to your evening plans as your child makes this transition.

Create a calming ritual that works for you and your child The timing and order of your soothing, calming activities will vary. You may choose to carry out all of your calming and connecting activities only in the room where your child is going to sleep. Or you may decide to complete some before moving into his "nest," and save one last favorite for when he's there. Then again, you may not move him into his sleeping space until he's completely relaxed. It's going to depend on your child.

If it works for your family, you can intersperse the calming activities with the tasks of getting ready for bed. Read a story together, take care of the business of getting on pajamas, leave your child to play quietly or listen to music while you care for another child, come back, give a massage, then lie together, talking.

Some children will only need one "connecting" activity with you, unless it's a bad night. Others will consistently need more. By having one activity, like massage or reading, that is easy to extend, you can eliminate the arguments. If, for example, on a particular night you sense that a battle is brewing, you can simply choose a book that is a bit longer, or massage a little more. By doing so, you can sensitively meet your child's needs yet remain consistent to the routine. And don't forget to let your child know what you are doing. You can say, "I think you are having trouble unwinding tonight. Would a little extra reading help?" Someday in the future, he will be able to tell you he needs that help instead of starting a fight with you.

The connection and calming activities are finished when your child's body is relaxed. Then it's time for the cue that the routine is finished, and it's time to slip into sleep.

Cue for sleep When Kristina was little, I always rubbed her back and sang *Hush, Little Baby* to her just before she went to sleep. Even during high school, there were nights when she was stressed, and my then seventeen-year-old, five-foot-ten daughter would plead, "Mom, please sing me my song." I'd sing to her then, and we would joke about who was going to sing to her after she left for college the next fall.

Imagine my surprise when a year later, as we were driving home after her first year in college, she turned to me and said, "Mom, my lullaby really came in handy this year." "It did?" I asked, rather incredulously. "Yes," she said. "During exams, when everyone was stressed, they'd say, 'Kris, sing us your song,' and I'd sing it. They really appreciated it. And when Anne's dad died, I sang it to her so she could sleep."

The cue for sleep is the last thing you do together with your child. Steven Porges at the University of Illinois Medical School in Chicago has found that changing your voice inflection but not yelling persuades the brain to shift from alert to calm. That's why singing a lullaby, turning on a favorite soothing CD, repeating a prayer or a chant, and even a simple *shhh, shhh, shhh* can help your child move across the line from alert to sleep.

Families whose children sleep well can always describe for me their final cue for sleep. "We say a prayer," Jacob told me, and then he looked a bit sheepish as he admitted, "We say, 'Night, night, sleep well, see you in the morning, sweet dreams, kiss, kiss, hug, hug, time to lay your head down.' Even our nine-year-old still does it." Find what works for you, a few endearing words that you are comfortable with becoming part of your family's history.

Your cue may also include turning on a fan, white noise machine, or night-light, tucking blankets in tightly, or turning off a light.

Switch to sleep It's at this moment, when all systems are aligned (body clock set, body relaxed) and your child has had a smidgen of practice, that he will allow you to step away while he adjusts his "lovie," rolls over, squirms a few minutes until he finds the right position, and falls asleep. All he needs from you is a promise to check on him again in five minutes.

If, however, you have the child who traditionally has shrieked bloody murder at the mere thought of you walking away, don't despair. He may surprise you with how quickly he falls asleep after following a routine that eases the tension in his body and works with his body clock. He may even be willing to venture a shot at letting you step away for a minute or two. But if he's not ready for that, it's just fine. While our final goal is for your child to pass through the switch to sleep independently, it does not mean abandoning him. We can help him get there with a little supportive practice.

Once you've given the cue that it's time for sleep, mentally check your child's body. If you note a lingering tension in her body, sharpness in her movements, or quick breathing, rub her back softly, saying to her, "You are safe. It's all right to go to sleep. I will stay with you." Encourage her to close her eyes, rub her forehead softly, moving your hand along her brow so that her eyes naturally close. "Shh, shh, shh," you may add, still rubbing softly.

When you sense a relaxing of the muscles, a slowing of the motion, and a deepening of her breathing, stop touching her, but stay near her. Her eyes may open. Remind her: "I am here. You are safe. It's OK to go to sleep." Sit or lie quietly by her, but do not touch her, unless she tenses again. If she does, stroke her soothingly. As you sit or lie by her, listen to her breathing; enjoy the softness of her breath. Watch the shadows play in the room, allow yourself to muse. Let go of your "to do" list for now, replacing it with a sense of tranquility. She will sense your peace, know it is safe, and allow herself to sleep.

As her breathing deepens, move from the bed to a chair next to it. Again, if she alerts, tell her you are here. She can sleep. If she needs it, and is old enough that you are not concerned about smothering, tuck an extra pillow tightly next to her body, "nesting" her into her sleeping space. Watch carefully, noting the position she moves into, the motion of her hands, the sensations she seeks, as she falls asleep.

The next morning, celebrate with her what a great sleeper she was and how beautifully she fell asleep. Ask her if tonight she may be ready to practice falling asleep with you stepping out of the room for thirty seconds. When she says "No," you can reply, "OK, I will stay with you until you fall asleep, but soon you'll be ready." And one day, she will.

If your child is an infant who is too young to understand your words and is a sensitive sleeper with a strong survivor drive, your child is NOT going to let you put him down until he is asleep.

We'll respect that, too. If he needs to be nursed or rocked until he's asleep, we'll do it, but we will do it in his sleeping space, with the lights off or low, and his cue music or fan running. Gradually, he will begin to associate these sensations with sleep, and he, too, will one day be ready to lie down to sleep. That's what Barb, a seasoned mom of three, discovered.

"I'd take six-month-old Nathan into his bedroom, turn off the lights, turn on the lullabies, and feed him his bottle and rock him. If I tried to lay him down before he was asleep, he'd shriek. He's really an intense and sensitive kid. He'd get so upset that it wasn't worth it. So I just let him finish his bottle until he went to sleep. Then, at about ten months, he was still awake after the bottle was empty, but I knew he was really tired and that if I let him get up again, he'd be overtired and I'd miss his 'window.' So I started the music, laid him in his crib, rubbed his back, patted his bottom, then stopped and stood there. I couldn't believe it. He flipped over, pulled his blanket to him, and, in a few minutes, went to sleep. Now he does it every night. He even turns on the music himself."

Ten months may feel like a very long time, but some babies are not ready to make the "switch" until they've gained a bit of maturity and practice. Just as you will carry him until he's ready to walk, and not think anything about it, you will assist him in moving into sleep. Eventually, he will get there. Throughout life, there will be times when more support is needed again, but with a routine in place, you can rely on it and fall back on it as a guide.

A month after his first success, Nathan was sleeping thirteen hours a night, uninterrupted. He had learned that he was safe—to sleep.

A routine also prevents the bedtime debates If your child is under three years of age, you'll follow this routine exactly, helping him to learn the steps. A clear routine helps you to sensitively but firmly "hold the line." When your child begs for another story, your routine will allow you to say to him, "You wish you could have another story, but one story is our rule." He won't like it, nor will he say, "Thanks, Dad, that's really what I needed to hear," but that's all right. Sometimes learning rules is hard work.

Create a picture planner It's also helpful if you create a picture planner of your routine so that you can point to the picture to show him the one-story rule. A visual planner includes five or six steps or pictures. It may include pajamas, snack, teeth, toilet, one story (or ten pages of a book), kiss, and lights out. When your child begs to diverge from the plan, you can simply ask, what does the picture planner tell us? If you've already read a story, you can point to that picture, and your child will know you are finished. Young children really do believe what they see, and it's also impossible to argue with a chart.

After talking with me, Paula and her five-year-old son Marty drew their bedtime routine plan—five separate pictures showing each step. Square one, pajamas; square two, two story books; square three, teeth and toileting; square four, good-night song and a prayer; square five, a kiss and a hug. The first night, everything was going smoothly, as together they checked the chart to see what came next. But after the kiss, there was a problem. Marty didn't stay in bed. "I think we need a tune-up for the plan," Paula said, when she called. "Did you try including a picture of staying in bed as the last square?" I asked. "*Hmmmmm!* No, we didn't." The next day, she sent me an e-mail. "I had Marty redraw his nighttime routine to include a stay-in-bed box. That night, he went to bed at seven o'clock and fell asleep at about seven-thirty, never getting out of his bed. A great night!"

A picture planner will help you to stay firm. If that's difficult for you to do, imagine for a moment that your child wants to play in the street. You probably would have no problem confidently and clearly communicating to your child that this ridiculous idea is not an option. You would stand tall; your voice firm but not shrill. The answer would be so obviously "no" that there would be no room for a fight.

You would not allow it, no matter how vehement your child's protests or even if all the neighborhood kids were dashing across the street at that very moment. Now apply that same confidence to your bedtime routine, because sound sleep is just as critical to his well-being.

Customizing the plan to fit your child

Every family's bedtime routine is going to look a little bit different. Experiment until you find the timing and strategies that work best for you and your child. You may also need to adapt it when tension levels are extra high, or a major event has disrupted your child's sense of security. These changes are so important that I'm dedicating the entire next chapter to them. But at this point, I'm going to assume that all has gone well and the children are snuggled sweetly in bed. Now, the question is, what about you?

Getting your own sleep needs met Before you jump to your "to do" list, consider using the extra time you now have in the evening to spend with your partner. And happy adults foster contented children, so don't forget to get yourself to bed, too. Dr. Norbert Schwarz, professor of psychology at University of Michigan, conducted a study with 909 women. He found that an extra hour of sleep had more of an impact on how the participants felt throughout the day than earning more money a year.

And three recent studies have shown the connection between lack of sleep and obesity. Participants who had only four hours of sleep had a 24 percent increase in hunger. The study found that participants' appetite for calorie-dense, high-carbohydrate foods like sweets, salty snacks, and starchy foods increased 33 to 45 percent.

So, if you'd like to be happier, manage your weight, and improve the quality of your adult relationships, go to bed yourself. Sleep is golden. Grab yours.

SAMPLE ROUTINES

Just in case you need it, here are a few examples from other families. The Benson family has four children: the four-year-old twins have napped. One four-year-old needs less sleep than the other. Sometimes one, sometimes both parents put children to bed. Goal: Time alone with each child before sleep.

	Four-year-old	**Four-year-old**	**Six-year-old**	**Eight-year-old**
Transition activity	7:30 Pajamas, teeth and toileting	7:30 Pajamas, teeth and toileting	7:30 Pajamas, teeth and toileting	7:30 Pajamas, teeth and toileting
Calming and connecting activities	7:40 Looking at books in bed	7:40 Playing quietly in bedroom	7:40 Reading with parent	7:40 Reading to self in bed
Cue			Song, prayer, kiss and hug	
Switch to sleep			7:50	
Calming and connecting	Parent reading and massage	Continue play or look at books in bed		Continue reading
Cue	Song, prayer, kiss and hug			
Switch to sleep	8:05			
Calming and connecting				Parent snuggling and talking with/ or reading
Cue				Song, prayer, kiss and hug
Switch to sleep				8:15
Calming and connecting		Reading with parent		
Cue		Song, prayer, kiss and hug		
Switch to sleep		8:30		

Check backs occur as needed.

The Nelson family has three children. One parent puts the children to bed. Goal: Make it simple.

Child	**Infant, 11 months**	**Four-year-old**	**Six-year-old**
Transition activity	7:30 Pajamas, diaper	7:30 Pajamas, teeth and toileting	7:30 Pajamas, teeth and toileting
Calm and Connect	7:40 Play on floor while parent reads to siblings	7:40 Snuggle up on couch with Mom and brother for reading	7:40 Snuggle up on couch with Mom and brother for reading
Cue	8:00 Rock and feed bottle, lay in crib, music, night-night-sleep-tight, kiss, hug, few pats	7:55 Into bedroom with brother, night-light on, prayer, kiss and hug	7:55 Into bedroom with brother, night-light on, prayer, kiss and hug
Switch to sleep	8:25	8:00	8:00

TWELVE

CUSTOMIZING THE BEDTIME ROUTINE TO FIT YOUR CHILD

"I'm lying here quietly, why aren't you?"

—Robert, father of three

The flame danced upon the logs, bright orange coals contrasting with blackened wood. An occasional snap announced the tossing of sparks into the dark cavity of the fireplace. I had brought the logs in earlier, piling them into my arms from the neatly ordered stack under the maple trees. Bright yellow leaves had filtered down upon me. Now I saw one lying on the hearth, a stowaway on the logs I'd brought inside.

The sun had slipped below the horizon, glowing red, as autumn skies do, and darkness had settled in. Supper finished, dishes washed, it was time to sit and enjoy the fire. Initially, I read, but then my eyes grew weary. I switched off the lights and sat watching the flames pirouette and leap across the stage. My muscles loosened, sleepiness crept upon me, easy, friendly, the thought of climbing into my bed a pleasure.

How easy it is to fall asleep when we set the stage, yet how challenging it is to do. Maybe that's why it's actually a gift to have a child who teaches us the importance of slowing down our lives, connecting with one another, and gradually transitioning into our night's repose.

When you have the sensitive sleeper who struggles to sleep unless the conditions are just right, it's often difficult to appreciate his role as the "emotional barometer" of the family. He absorbs the stimulation, reacts to the hectic pace, and picks up the stress of the environment, soaking it all up into his body, winding tighter and tighter until he announces, "I don't need sleep. I sleep with my eyes open!" And, despite our best efforts, his eyes do remain open, screams erupt, and, once again, this child takes twice as long as the other kids combined to get to sleep. But even the sensitive sleeper can become a "good sleeper." It requires finely tuning our system and valuing that this child, like a canary in the mine, is the family's warning system, unable to sleep when our lives are out of balance.

CUSTOMIZING THE PLAN FOR TEMPERAMENTAL DIFFERENCES

It's not your imagination when you feel that you are working harder with this child than your others. Or that your lights are on in the middle of the night more frequently than those of your neighbors. It's a fact. It takes more: consistency in your routine, careful monitoring of stimulation, connecting and calming activities, and time to help this child's brain "switch" to sleep. And it doesn't stop there. You also have to observe more keenly and work harder to precisely catch his "window."

I know you may wonder why you have to work this hard. Or why this child was sent to you instead of your sister, who really deserved him. But what you will find as you work with him is that when he is well rested, he can also bring you *more* pleasure. His passion for life, warm sensitivity and energy are delightful and well worth the effort. It begins by predicting the unique challenges you face as the parent of a child who is "more."

Predicting the trouble spots

Think about it: what do you fight about every single night? Is the feel of your daughter's pajamas an issue? Does your son tickle and touch your back when you try to lie down with him? Will the youngest "dive-bomb" the group when you try to read to the others? Is there always someone who begs for one more story, or a few more minutes to finish a project? Does your daughter play with the toys in her room instead of getting on her pajamas when you've gone off to put the baby to bed? These behaviors are actually linked to your child's temperament. If you can predict them, you can plan for success.

If your child is intense

Intense kids are the "Corvette" kids. Their engine runs high and fast. While some children, who are less intense, may take or leave a calming activity, you can predict that your intense children will crave every single one. They need more time to slow down and put on the brakes. That's why it is so essential to catch them early and help them take it down a notch.

It wasn't until later that Lisa learned that while she was out with friends, her two-year-old daughter Emma was also having a good time. In fact, at nine-thirty p.m., when she had demanded ice cream, her grandparents happily loaded her into the car and treated her to a chocolate sundae. But that didn't slow Emma down. At six a.m. the next morning, she was up and streaking stark naked across the front yard, with her mother in pursuit, Emma squealing in delight.

Wired, wild, and shrieking. The entire day continued that way until nine-thirty p.m., when Emma fell apart. Exhausted but unable to sleep, she kicked when her mother attempted to rock her, squirmed out of her arms and stood three feet away, glaring, challenging her mother to catch her if she could.

Catch and connect The moment you realize that your child is in a state of tense tired, whether it's at six o'clock a.m., the middle of the afternoon, or at dinner time, it is essential that you catch your intense child's tension and take steps to ease it before it overwhelms her. Stop what you are doing, do not make any more demands, and, instead, attempt to calm. Don't worry that the clock is ticking and you have other things to get done or an event to attend. By slowing down to connect and calm, you are going to save yourself a great deal of time and avoid a brawl of power struggles.

It's not easy to do, especially when your child has just demonstrated her tense energy and exhaustion by announcing, in a very unpleasant tone of voice, that she hates everything you have just prepared for dinner. But the best response at that moment may be to simply give her the look that says "stop," and then, if she is open to it, hold her quietly for a few moments. As you do, think about what soothes and calms her. Does she respond well to a warm hug? If you simply shut things down for a moment—stop talking, dim the lights, turn off electronics, or allow her to take a break—does it help? What about getting up and moving? Or, perhaps, she responds better if you try a bit of distraction by turning on soothing music or reading together.

Work together When you stop to calm your child, you are not ignoring the poor behavior, you are simply waiting for the "teachable moment." Until she is calm, she cannot hear you or work with you. You will go back later, once she is composed, to teach her that even when she is tired, she must still treat others respectfully, but that lesson must wait until she is not in a state of tense tired. In the meantime, you are working with her, helping to ease the tension in her body, and bringing her back across the line to calm tired, so that she will be able to relax and get the rest she needs.

It may be later that evening or even the next day before you can say to her, "I think you were very tired at dinnertime and, as a result, nothing looked good to you. When you are feeling that way, you can say, 'I don't care for any, thank you.' Or you can say, 'Thanks for making dinner, but I'm too tired to eat.' " By waiting until she can hear you, you'll make your words more effective. She wants to be successful. She doesn't like to get into trouble, but sometimes she struggles when she's too tired.

Plan tension busters If you can predict that she will frequently be tense after school, or at dinnertime, plan a calming activity, like drawing, reading, or exercising right after school and BEFORE dinner, so that you do not move into your evening fighting. If this occurs frequently, consider enrolling her in a yoga class. Studies show that the regular practice of yoga leads to better sleep.

"I absolutely dreaded dinner and bedtime at our house," LaRyn told me. "We've got four kids. Dinner was always a disaster, and from there we moved right into the bedtime battles. But after realizing we could predict it, we instituted 'quiet time' before dinner. Now, if I'm busy in the kitchen, they color or draw at the table. If dinner is in the oven, I sit down and read aloud to them. It has TOTALLY changed mealtime. Now they sit, take turns talking, and eat. We move into the evening and bedtime calm. I don't feel panicked and exhausted. I think they feel better and sense that I do, too. It makes such a positive difference at bedtime."

Managing the intensity throughout the day, especially in the late afternoon and evening, is critical if the intense child's body is going to be ready for sleep.

Catch the "window" An overtired intense child courts disaster. Observe closely; finely tune your observations until you KNOW your child's cues that she's moving toward her "window." Once you are aware of the cues, teach them to your child. Point out to her how her voice tones change, saying, "When I hear that tone, it makes me think you are getting very tired." It's likely that she will retort, "I'm not tired!" That's all right. You're helping her to become aware. Someday in the future, she will accept and use this information herself. Most important, sense when to move her bedtime earlier, so that she does not become overtired. Even if it means changing plans, it will save you both a great deal of frustration and turmoil.

Continue taking it down a notch at bedtime If, as you move your child to her sleeping site, she refuses to get into bed, won't stop touching or tickling you, or otherwise won't settle down, don't fight it. Remind yourself that she is having trouble slowing her body down. She is not trying to push your buttons.

What you can do is NOT add to the intensity. Instead, move away from her, and sit and rock for a moment, or simply lie on her bed to calm yourself. Don't interact, and don't chase her. Your calm demeanor will begin to quiet her.

"Sometimes what calms him," Scott told me, "is for me to simply lie down and pretend that I'm asleep. I breathe slowly and deeply. He's running around the room, but gradually he starts to slow down, then moves in toward me, and somehow my breathing seems to slow his and he calms down."

After a few minutes, if she's young, pick her up, wind your arms around her, and rock quickly for a few minutes. When she squirms in your arms, let her go again. Once more, sit calmly and wait a minute or two. Then gather her in and rock again. When her body begins to relax, lay her down on her bed. If she sits up, gently lay her back down again, stroking her back and face, telling her she can let her body slow down. It's time to stop. When she says she wants to get up, respond with empathy. "You wish you could get up, but it's time for rest. Your body is very tired." Following her cues, adjust your touch to firm or light. You may even lie next to her, with your arm across her body, helping her to shut down.

If she's older, invite her to take deep breaths, to relax her toes, her legs, and arms, or allow her to read on her own for a little while. Keep that image of winding down in your mind. If it feels like she's inviting you to chase her, or talk to her, decline the invitation.

Stay cool, bring it down, and you'll find that even your intense child will sleep. Once you sense that she is completely relaxed, try telling her you will check on her again in two minutes. If the mere thought of your departure brings back the tension, recognize that today is not a day for practicing independence. Once life has settled down a bit, you'll try again, but on this night, plan to stay with her as she slips into sleep.

If you find that she needs you close, but revs up or can't seem to stop if you're lying next to her, try having her lie with her head at the foot of the bed while you sit with your back to the headboard. You will now be toe to toe, still close but it will be easier for her to settle.

Once your child does fall asleep, give yourself a big pat on the back. You did it. You stayed calm. You can promise yourself that tomorrow you'll make an extra effort to help her manage the pace of her day, reduce the tension as it occurs, and to hit her "window" for sleep. It will be a good reminder that while you can't "make her sleep," you can plan for her success.

And if, by chance, you've been blessed with two live wires, you may find it easier to lay their mattresses on the floor and sit between them, stroking each with one hand. Or you may choose to separate them and work with one and then the other.

If this type of sensitive care goes against your grain, you are not alone. Brad was a dad who e-mailed me after one of my workshops: "I'm one of those people who grew up with the dad who said, 'You're going to do it because I said so.' That's the kind of person I was, too, but now, instead of fighting, we're connecting, and he's sleeping."

Connecting calms your child. A relaxed body slips into sleep.

If your child is sensitive

It is sensations and emotions that "alert" the sensitive child. That's why it is so important to create an atmosphere that is quiet and peaceful rather than loud and upsetting.

Give your child words If you can predict that putting on pajamas will be a battle each night, before you start getting them on, ask your child what her body is feeling tonight. Is she hot or cold? It's especially important that her hands and feet are the right temperature. Does she want a firm waistband, a loose one, or would she prefer a nightgown? Remember, this is the child who can tell the difference between 100 percent cotton and 50 percent cotton, 50 percent polyester sheets.

Teach her to say, "This doesn't feel quite right. May we please find a different one?" Then she'll have words to use instead of resorting to screeching and shrieking when something is disconcerting to her.

If your child is an infant or a toddler, talk to her as you make these judgments. When she squirms or fusses as you put on her pajamas, say to her, "I see you squirming. Is this scratchy? Let's find a different one." She will hear your words, and one day surprise you by saying, "Scratchy—no like." It's so much easier to keep your cool when everyone is using words instead of yelling.

Talk about emotions It's the sensitive child who is keenly attuned to the emotions around him. This is the child who asks you why you are upset before you realize that you are. Time to talk and ease the emotions of the day is essential for him and needs to be part of his evening routine. If he is old enough, you may introduce a worry journal. This is a journal in which you help your child to record his concerns. Then shut the journal, and put it away in a drawer, letting him know that he can let go of the concerns of the day.

Extra reassurance that it is safe for him to sleep is often just the right tonic for the sensitive child. You can help him feel comfortable by letting him know what you will be doing while he's falling asleep, or promise to check back in on him in five minutes. A simple, "I'm going to go call Grandma, or switch around the laundry, and then come back to check on you," can make a world of difference to him. Remind him that it is all right for him to sleep.

Because he does pick up your tension, it's important to affirm how perceptive he is while, at the same time, assuring him that the adults in the family will take responsibility for solving problems. A simple phrase, such as "You're correct, Mom is upset, but you can sleep. Mom is a good problem-solver, and she'll figure out what to do," will help him relax. Do take care of yourself. Even in sleep, this child remains aware of stress in the environment.

This is also the child who needs to know he's not the only one who feels the way he does. You might try telling him something like "You and Dad are both very sensitive to noises and smells. Dad needs to have quiet, too." Or, "You really can smell the new detergent on the sheets. Grandma always notices that, too."

Help her find the right touch Rather than soothing him, the bath may turn your sensitive child into a "wild man." But watch carefully what happens thirty minutes later. Sometimes splashing around in the tub with the pressure of the water against his muscles and joints is exactly what he needs. Just make sure that it is early enough for him to release all that energy from his system before you expect him to fall asleep. Often a bath before dinner is most effective for this child.

The same is true with wrestling. Some highly sensitive children actually benefit from the deep pressure of those bear hugs—but again, it usually needs to be EARLIER in the evening. Watch carefully, and he will show you when and what type of touch and sensations he needs to relax. If he does get wired, and it's close to bedtime, give him a few weighty books to carry upstairs, and then read to him while he's sitting in a beanbag chair with a heavy pillow on his lap. Use low light while reading. These activities create pressure and give the muscles feedback to relax. Plan in the future to make sure that, during the day, he has many opportunities to carry, pull, climb, pump, kick, and push, so that at night his muscles are ready for rest. Strictly limit his exposure to video, because of the stimulation it creates and the lack of exercise it fosters.

Provide white noise Sounds easily alert the highly sensitive child, so you will want to be sure to add a sound buffer to his sleep environment. A fan, aquarium, humidifier, or sound machine may effectively block offensive noises. The key is to find the right steady drone that allows him to sleep through the household noises. Otherwise, he will hear all of the sounds in the environment and stay in a state of heightened arousal, never able to relax and sleep deeply.

Limit visual clutter One day my daughter and I walked into a local bakery. I knew immediately that I wanted something chocolate, and as soon as I spotted a chocolate chocolate-chip muffin, my selection was made. My daughter walked out without purchasing anything, even though it had been her idea to stop there. When I asked her why, she explained that the number of choices had overwhelmed her. She proceeded to describe to me the ginger, banana, banana walnut, carrot, bran, blueberry, lemon, cranberry, as well as several other muffins she'd seen. The list went on and on, and I wasn't surprised by the fact that she had found the number overpowering. But what did surprise me was that I hadn't even noticed all of the choices. Once I saw the chocolate one, I was finished, and stopped perusing the shelves. Kristina, who is much more sensitive to her environment than I, could not.

Highly sensitive individuals sort visual stimuli, which is why it's essential to strip down your sensitive child's room and get rid of the visual clutter. Look carefully. Is there equipment or toys that are begging to be played with? Does your child sleep better if the walls are bare?

If your child is slow to adapt

"When I tell my son to come inside for bedtime, he revs up. He doesn't want to stop and brush his teeth." "My daughter argues whenever we ask her to get ready for bed." These are common challenges faced by parents whose child is slow to adapt. Transitions and surprises trigger the slow-to-adapt child. What's most important to remember is that he's not just being stubborn. Physiologically, it's more challenging for his brain to "switch" from one thing to another.

It's the slow-to-adapt child, more than any other, who needs warning that bedtime is approaching, so that he has time for closure and a predictable, consistent nightly routine. A haphazard, random schedule or rushing can inadvertently lead you into a nightmare of arguments and debates.

Be aware of your own adaptability "I don't want to go to bed. I'm not tired!" The slow-to-adapt child rarely greets the announcement of bedtime with pleasure. Instead, you can expect protests. That's why it's so important to be aware of your own adaptability. If you are highly adaptable and easily shift from one thing to another, your child's complaints may lead you to avoid the fuss and keep adapting the routine to accommodate him. The trouble is that the more alterations you make, the greater the likelihood that you'll end up with a slow-to-adapt child so exhausted that he has completely lost any ability to shift. By maintaining a consistent plan, you make it much easier for both you and your child.

Consolidate transitions Karen walked into class looking a bit frayed. I turned to her with a questioning gaze. "Looking at books in their beds is working fabulously," she said. "They are quiet and relaxed. But as soon as I come in to give the signal, they seem to get agitated and start stalling. One night, I got so panicky about going in to give the signal that I never did. I just let them keep reading until they fell asleep. What would you think about doing the signal of good-night prayer and kiss, and then letting them decide when they are tired and ready for sleep? Or am I just being a wimp?"

What Karen had discovered is that slow-to-adapt children can be aroused every time there is a transition. As you create your routine, minimize the shifts. Once you are in the bathroom, take care of all business there. If you've gone upstairs to bed, avoid coming back downstairs for a snack. Conduct all calming activities where your child will sleep or on his bed, so there isn't a change of scene when it's time to go to sleep. And consider consolidating all of the activities that you do with your child into one block, so that you are not going in and out of his sleeping space. The key is to find what works for you and your child, by minimizing the transitions that alert him.

Manage the callbacks The slow-to-adapt child is the king of callbacks. "I need another drink of water." "I have to go to the bathroom." "Can't we read one more story?" Expect it and plan for it. Consider a "callback ticket" or two. Let your child know that he has one "callback ticket." He can decide whether or not he wishes to use it. What you'll discover is that he'll often save it. When he does, compliment him the next morning on how well he settled himself and slept. By making the number of callbacks concrete, you meet his needs but also clarify the limit. Do your best to avoid ambiguous limits, such as "I'll lie here *for a while.* Or "I'll come back for the next *ten minutes.*" Unless you are using a visual timer that lets him see the approaching end time, the vagueness can actually create anxiety for him and make it tougher for him to settle down.

If you know your routine is consistent and limits are clear, yet your child is still calling you back, it's likely to be an indication that something is amiss. Has his "window" for sleep moved for some reason? Has there been a disruption of the routine earlier, which may have upset his body clock? Is there something that may be leaving him feeling a bit more anxious or tense? Slow down; take a few extra minutes as you reassure him that it's all right to sleep, stroke his face, let him know what you'll be doing, and ease that tension.

If your child's body rhythms are irregular or unpredictable

Despite his protests, or his limited cues that he's hitting his "window," do your best to keep him in the routine seven days a week. The improvement in his behavior during the day will be well worth the extra energy it takes at night to head him off to bed. The predictability of that bedtime is essential. This is the child who needs to hear from you, "You may not feel tired right now, but let's do our routine, and see how you feel." Or, "I know you'd like to stay up, but it's very important that you get your sleep."

If your child is energetic

Winding down is very challenging for high-energy children. They have so much to do! When your high-energy child is having difficulty settling, remember, instead of thinking why is he like this, recognize that he's having trouble slowing his body down, and consciously think of ways to help him channel and release his energy.

Plan for the energy Maggie found a great solution. "Four-year-old Evan is our 'dive bomber.' He wouldn't sit with us and read; instead, he'd dive into the group. Now I go into his room and put all the loud, enticing toys, like the trucks and tools, into the closet. Only then do I dim the lights, pull out the books, and bring him in. He's a high-energy child, so it's hard for him to just sit and listen. I started giving him finger puppets. He likes to wiggle his fingers in them and pull them off and on. I think it's very soothing to him."

If you have more than one high-energy child, you may also find it helpful to read to them while lying on the floor, using a flashlight. It gives them space to wiggle without bothering one another, and the dim light helps to cue their brain that it is time for sleep.

When you understand how difficult it is for the high-energy child to slow down and sleep, it gives you much more patience. That's what Eve discovered.

"I thought of you a couple of nights ago, when I was putting Tyler to bed," she wrote to me. "He was doing handstands and somersaults in bed after lights out. I was trying to figure out why. I think it might have been because it rained all day and he'd played quietly with his trains. There wasn't much running around, and he's the child who really needs that exercise. It was also later than normal, so he was overtired. Just knowing that he wasn't trying to drive me nuts helped me keep my cool."

Do what fits your child What you begin to recognize as you work with your child is that every child is different. The commonalities are the need to cue for sleep by maintaining a regular routine, calming and connecting to reduce tension, and then adjusting to fit your child. How you do it is going to vary even within your own family. What's most important is that your child goes to sleep feeling calm and safe, and that your day has ended peacefully. Understanding temperament helps you to be successful.

When I walked into the center, Tammy came running toward me, proclaiming, "I could kiss you on the lips!" Now, I'm from Minnesota, and I like my kisses, but this was a bit more than I was ready for. I smiled, and before she could grab me, I asked, "What's happening?" "My son is sleeping so well." She laughed with glee. "What did you do?" "Well, after we talked about how sensitive and slow to adapt he was, I went in and took everything off the walls in his room. His window had been bare, so I hung a sheet to cover it. It's funny, because he would sometimes put his arm over his eyes when he slept, but I never picked up on the fact that he needed more darkness.

"After that, I went and got his rocker from the family room and put that in his bedroom, too, so when I nurse and rock him before bed, we're in his room. That way, the environment stays the same when he goes to sleep. Then I found a nice, soft baby blanket and his bunny with silk ears. I laid the blanket over my body, and then placed him against it as I rocked him.

"Squeezing my finger has been his lovie. It's been really sweet, but I was ready to change it. When he started to squeeze my finger, I wrapped the silk bunny ear around my finger and then let him squeeze.

"When he was almost asleep, I laid him on the blanket and put the bunny right next to his hand. I sat back down in the rocker and simply said to him, 'I'm here.' Initially, he popped right back up, but then he lay down again. I told him I'd come back and check on him. And then I walked out. I heard him roll over a couple of times, but then he just went to sleep. And he's sleeping. He's not waking up in the middle of the night. For the first time in his life, he slept from seven-thirty p.m. until seven o'clock a.m. It's been surprising how quickly it happened and he never, ever had any crying. I love you!" she shouted, and kissed me. Temperament is the tool that helps you fine-tune your bedtime routine to fit your child.

When your child *still* doesn't sleep

If, despite your sensitive, responsive care, your child screams bloody murder with the mere mention of bedtime, take the fight as information. It doesn't mean that your original plan was bad or that you are failing, but that something has shifted. There's a feeling or need that requires attention. It may be that your child is feeling too unsafe to sleep.

Perhaps there's been not just one incident, but a pileup of life events, like a new baby, a move, changing beds, Mom's return to work, Dad's traveling more, holidays, guests in the house, a new teacher, and/or an illness that shakes your child's sense of well-being. Not infrequently, if you have innocently followed the "let the child cry himself to sleep" method in the past, and especially if your child gagged and vomited, he may be feeling anxious at bedtime rather than relaxed. If this is the case, we have to help your child rebuild trust that his world is safe and it's all right for him to sleep. That means that if you've been frustrated with your child's behavior and have threatened to leave him if he didn't change, or have resorted to spanking, STOP! Your child needs you to choose to connect.

REBUILDING TRUST

When a child is feeling vulnerable, she will require more help falling asleep. We often worry that we are starting a bad habit when we support our children during tough times. What we have to remember is that it is a habit, and habits can be changed. When the stress has been reduced, and once again your child is feeling safe and secure, we will begin practicing switching to sleep independently once again, and, some day soon, she will be able to do it.

Allow time

If your child's sense of security and trust has been significantly disrupted, expect it to take time, often four to six months, to restore. It is ESSENTIAL that you do it, however, for your child's mental health and well-being. Find ways to support yourself, seek counseling, work with your partner, but REBUILD trust with your child. As you work with her, remind yourself that this is a battle with her brain, not you. Her brain is telling her to stay on alert. We want to calm it with reassurance.

Recognize your child's fears

Children tell you with their behavior that they are feeling anxious and afraid. Unfortunately, that behavior is often rather nasty and difficult to face at night when you're tired, too. Often, you'll see fight-or-flight behaviors like six-year-old Ryan's. "He kicks and screams, and then he starts demanding, 'When is the sun going down? I'm not finished with my day yet. I'm not ready to go to bed.' " Or, perhaps, your child will shut down like five-year-old Mia. "She'll lie on the floor. Some days she'll sit there and scream." And, then again, your child may demand to be "gathered in," like three-year-old Latisha. "She won't leave my side and won't go to the bathroom unless I am with her and holding her hand. If I'm helping her sister, she'll get aggressive. She wants it to be only us, no one else. She chases me if I walk away. If I try to force her to do something, it's only worse." These behaviors reflect a child who is deeply in the "red zone" of tense energy and tense tired. It's true the behavior is unacceptable, but first we have to calm her.

Reduce the demands and number of choices A child who is in the "red zone," can't think and doesn't function well. Often, she'll even become anxious before the bedtime routine begins. This is the time to prepare her, reduce the demands, and lower your expectations. If you know it will be a fight to get her up the stairs, offer to carry her. Before she starts to throw a fit, merely ask her, would you like to be carried? If she says yes, do it, and tell her that one day soon she won't need it, but now she does, and you will. If she's too heavy for you to carry, use distraction. Give her a job, like holding the door open at the top of the stairs, or taking up the basket of clothes you'll fold while she's winding down. Reduce the debates by eliminating the choices. Use the SAME books, pajamas, blankets, bedtime snacks, and songs every night. This won't be forever, it's just for now. Assist her in brushing her teeth, putting on pajamas, and getting ready.

Stay with her When your child is experiencing distress, it is not a time to teach independence. Your child needs to know you will stay with her. You may choose to let her sleep with you, or in your room. Or you may realize that you will not sleep well if she's there, so you will choose another room. If she's not sleeping with you or in your room, give her permission to come to you during the night, or reassure her that you will come to her when she calls. Then do it. In order to relax and sleep, she has to know you are available. I know this idea may send chills down your back, thinking that your child will be showing up in your room every night, but what I've found is that once a child knows that he can come and get into bed with you, or slide into a sleeping bag on your floor, he doesn't. Because knowing that he can relaxes him enough to sleep through the night.

Mike and Mindy found this reassurance was what their five-year-old son needed. "We were at our wits' end. But tonight, as you advised, Mike lay with him until he was asleep, and didn't even suggest that he do any of it on his own. He fell asleep after about an hour (it's been taking two or more) and slept through the night. This hasn't happened in weeks! He had a GREAT day and, of course, so did we."

Maintain a consistent, predictable routine Despite the fact that your child is resisting bedtime, keep the timing and steps of your routine consistent. What you will discover is that when it is predictable and includes calming and connecting activities, it will actually create a buffer and help your child get the sleep he needs to cope with the stress.

Establish clear limits An anxious child is so wired that she can't stop talking. She wants you to hold her, but then pushes you away or even hits you. Establishing clear limits and gently but firmly enforcing them actually helps her to settle. "Four-year-old Elli has two books she can look at while lying in bed," her dad told me. "If she continues to talk, I warn her, 'You get three chances to stop talking and slow down. This is your first warning. If you talk two more times, your books will be done.' It really seems to work, to give her three chances."

At this point, many children will simply finish their books, set them on the nightstand, and allow you to soothe them. Sometimes, however, they need more. That was true for three-year-old Mathew.

"After we finished reading, I'd hold and soothe him. I would remind him that I would stay. He was not alone. Initially, even that wasn't enough. He'd push me away, but didn't want me to leave. Then he'd start in with the little things. 'I missed a page in my book.' Or 'This isn't the right stuffed animal.' Or 'I need a drink of water.' It was as though he was desperately fighting sleep. I kept reminding him that he was safe, but he'd squirm and whimper and then, five to ten minutes later, start screaming."

If this is your child, remind yourself once again, you are not doing something wrong. He is highly stressed, and his brain is desperately trying to keep him awake and alert to meet the threat. You don't want his sleeping room to be associated with screaming, and if others are sleeping in the room, his cries will disturb them. Take him out of the room.

He will continue to cry. Stay with him. Don't talk to him. If he hits himself, or bangs his head on the floor, gently stop him. Let him go through the cycle a few minutes, and then tell him, "When you are ready, you can sit with me. I will hold you. I am waiting." If he insists he needs something from his sleeping room, let him know you will get it as soon as he calms down. After ten or fifteen minutes, a child will usually begin to calm. Hold him. Stroke his arms and feet. You may even use a soft brush to gently rub his extremities. Don't talk. Talking can fuel the intensity. If it is acceptable to him, place a light blanket over his eyes or head, to help him shut down. Once he is calm, take him back to his sleeping room.

Get another adult to back you up Depending on the severity of your child's distress, he may or may not fall asleep. If he consistently starts to scream again, or hits and scratches you, it's time to get another adult to work with you. The adult needs to be someone your child knows well—your partner, a grandparent, friend or neighbor, someone who has cared for your child in the past.

Ask this adult to gently remove your child from the sleeping room and to stay with her. It's likely that your child will scream for you. When she does, this adult can say, "I know you want your dad. When you are calm and won't hurt him, I will take you back to your sleeping room and you can have him." Once the child is calm, this adult then takes him back, comforts him in his "nest," and then lets you (Dad or Mom, whomever the child wanted) take over again. If he cries and hits again, you can say, "If you are screaming and hitting, you are telling me that you don't want me. Are you choosing to go with Grandma (the other adult) or would you like to stay with me? If you want to stay, you must stop."

This is not easy to do. It can be incredibly frustrating, knowing and predicting that your child will have to go through this cycle. You hope that today will be the day she will be able to calm her body and go to sleep. But then you can feel her building and know it's going to happen. The intensity of it can be frightening. And you may feel horrible for your child and what happened that is making it so difficult for her to sleep. Hang in there!

It took four months. But today, four-year-old Emma is once again easily falling asleep. She can now wait while her dad soothes her brother, and even allows her favorite sitter to put her to bed. The steps were incremental. Sometimes there were two weeks in a row of success, only to be followed by a backslide. During the rebuilding of trust, Emma's parents stayed with her, very gradually introducing mere moments of separation as she fell asleep. She knew that if she awoke in the middle of the night, someone would immediately respond to her. But now, thanks to a significant effort by her parents to connect and calm, Emma has found that her world is safe enough for her to sleep.

Consider the possibility of medical issues

If you find tears welling in your eyes at this point, realizing that you have done all of these things and still your child is not sleeping, it's time to call your doctor. Difficulty falling asleep and staying asleep can also signify underlying medical issues. If possible, ask for a referral to a sleep center in your area, which has specialists working with children. This is especially true if your child snores, which can be an indication of sleep apnea. A child with sleep apnea may be arousing as frequently as one hundred times a night, which would leave him exhausted the next day despite the time spent in bed "asleep."

Sleep problems can also indicate neurological and developmental issues for your child. Early intervention is critical. You are not alone. Help is available. There are excellent sleep centers all over the world.

THIRTEEN

NIGHT WAKING, NIGHT TERRORS, AND NIGHTMARES

Quieting the screams in the night

"I know she loves me and is just checking in, but sometimes I wonder, does she have to love me this much?"

—Sue, mother of two

The ringing of the telephone jolted me from sleep. I glanced at the clock, noting that it was three o'clock a.m. My heart lurched when I answered, realizing it was my daughter, Kristina, and she was sobbing. "What's wrong?" I gasped, as my heart raced. Choking on her sobs, she exclaimed: "It's my ear! It really hurts! I think I have an ear infection!" Kristina, a college freshman, was in her dormitory, a thousand miles away from home.

I sighed deeply, relieved yet feeling awful that there was so little I could do from such a distance and at this time of the night. It seemed that just having someone to talk to helped a bit. Unable to think of anything else, I asked her what she'd done. Her efforts to ease her pain had been thorough, yet she was still in agony. "You've done a great job," I replied, "but I think you'll have to seek medical care." "But I don't have a car and my resident advisor is gone and my roommate went home for the weekend." "You can call your brother," I suggested, since he was attending the same university and had a car. "But Mom," she wailed. "It's the middle of the night—I can't wake him up."

I couldn't help it, I started to laugh.

The reality is that from the moment our children arrive, our lives are changed. No longer can we count on an uninterrupted night's sleep. Whether we wish it or not, there will be times when they will awaken us in the night—even after they've moved out.

While the idea that children should sleep through the night has been pervasive in our society, it is a myth. No one—not infants, children, adolescents, or adults—always sleeps through the night. It is an expectation that has caused a great deal of frustration and guilt, and has left millions of young children needlessly screaming in the night. It is normal for children to awaken sometimes. We can relish the times when all is going well and everyone is sleeping, knowing and expecting that there will be times when they won't. When they don't, we can make it better. We cannot make it perfect.

Waking at night is often the result of what's happened during the day that has left your child in a state of alert. What we can do is choose to adjust our perspective AND carefully examine and change the things we are doing during the day that may be inadvertently creating more disruptions. Instead of being upset when our child awakens, we need to recognize that those disruptions may be a red flag that something is up. Through empathy and understanding, we can select the most effective techniques to help everyone get more sleep.

Why we wake

Sleep is often viewed as one continuous inert state. But sleep is actually a complicated process, during which the body and brain cycle through periods of activity and inactivity. Understanding this process can help you understand how easily it can be disrupted.

In chapter 8, I described for you the circadian rhythm, the twenty-five-hour cycle that tells the body when to be asleep and when to be awake. In addition to the circadian rhythm, we also have an ultradian rhythm. This clock controls the sleep cycles we pass through during sleep.

Very young infants go through stages of active, quiet, and a transitional stage of sleep that the scientists call "indeterminate sleep." In the first three months, infants' sleep is divided equally between active and quiet sleep, with a very small portion of indeterminate. By about six months of age, the sleep stages have developed into the two phases—of REM and non-REM sleep—that will remain with your child throughout life.

During the night, if you are three years or older, you will experience five to six sleep cycles, each lasting 90–110 minutes. Children six months to three years of age experience cycles that are 60–75 minutes in length. Moving through each cycle, you experience different types of sleep, including quiet, non-REM sleep, which consists of both light sleep and very deep sleep, and active, or REM, sleep, during which you experience most of your dreams. It's at the beginning and end of these cycles, and as you shift from one type of sleep to another, that you are most vulnerable to awakening. That's why your child often awakens every one and a half hours to two hours. On a good night, you will not be aware that your child is moving through the cycles. He will easily shift from one stage of sleep to another. But on a bad night, he will experience disruptions, awaken, and call for you to help him go back to sleep. Fortunately, we now know that these awakenings are not willful behavior, nor are they an intentional plot on the part of your child to torture you. Instead, the awakenings are a reflection of what is happening in his brain and body. Once again, the same culprits that make it challenging to fall asleep also make it more likely to awaken in the night:

1. Tension—high tension blocks deep sleep
2. Time—a disrupted body clock results in more frequent awakenings
3. Temperament—some individuals by their very nature are more prone to arousing

You don't have to be fighting with your child in the middle of the night. An awareness of these common culprits allows you to make decisions and take actions during the day that can help minimize the number and frequency of awakenings.

TENSION BLOCKS DEEP SLEEP

A study completed by researchers at the University of Pittsburgh reports that stress may disrupt the natural rhythms of the body's nervous system during various stages of sleep. The study found that stressed sleepers wake up more often and have fewer episodes of deep sleep. Stressed individuals experience changes in heart-rate patterns during REM and non-REM sleep.

If your child has been sleeping well and suddenly stops, or the stress level is high in your family and your child is awakening frequently during the night, the culprit may well be tension. Even if the troubling event occurred six to twelve months ago, its repercussions may just be appearing now. If that's the case, you can begin to reduce the impact by managing your own tension so that it doesn't spill over to the kids.

Adjust your expectations to minimize tension

Feeling guilty or inept creates tension. If you believe that your child is always supposed to sleep through the night, it will upset you when he doesn't. That was Sarah's experience. "Christopher pulls his blankets down the hall and climbs into bed with us. I can go back to sleep when he does, but there's always the guilt. I can't help thinking that everyone else has kids who are staying in their beds. I think they must have it easier, and I wonder if they're up at night or if we're the only ones."

If, however, you recognize that awakenings are normal and can provide you with information about what's happening inside of your child's body and brain, your perception changes, and your blood pressure goes down. Fortunately, there are now research studies that confirm: it's actually the *norm* to awaken during the night.

In a study completed by Antonio Beltramini, M.D., from Cornell University Medical College, it was reported that 95 percent of the children were described by their parents as regularly crying or calling out during the night as frequently as once a week, at some point during the first five years of life. Seventy percent of the children were noted to regularly awaken one or more times a night.

Scientists have also documented that infant boys awaken more frequently than girls, and children who are temperamentally more intense will awaken more often than those who are not.

Historically, school-age children have been known as the "best" sleepers, but that has been changing, as the pace of their lives has increased. The resulting tension and disrupted body clock have taken their toll.

And it's not just the kids. About one out of four adults who share a bed with someone report that their bed partner's sleep problems interfere with their own sleep. Forty-seven percent reported losing at least three hours of sleep per week, and 23 percent reported losing five or more hours. The leading reasons were snoring, tossing and turning, insomnia, and hogging the mattress and covers.

Recognizing that awakenings are normal and something to manage, rather than a problem that you have failed to fix, leads to very different expectations.

I asked a couple who had recently immigrated to the United States from Cameroon, West Africa, when they would expect a child to sleep through the night unattended. They answered, three. Uncertain what they meant, I attempted to clarify their answer. "Three months?" "No, three years," they replied. "Until then, children wake and need to eat."

In the days of Benjamin Franklin, people went to bed when it was dark and awoke with sunlight. That meant they often spent twelve hours in bed. They didn't expect to be sleeping the entire time. So, it wasn't unusual for them to awaken at three or four in the morning and lie wide awake, thinking. Benjamin Franklin called it his muse time, and valued it as a time of high creativity. He allowed himself to muse, and then returned to sleep.

When your child awakens in the night, know that other families also cope with night awakenings. Take a deep breath; remind yourself that there is a reason that your child's body and brain are disrupted on this night, and that it's probably due to high tension, a disruption of his body clock, or his temperament. You can take steps to make it better. This perspective will allow you to comfort your child and fall back to sleep more easily yourself, because you are not worried that you are a failure or that your child is being mean to you. We all like and need our sleep, so, let's take a look at some helpful strategies.

Mark your calendar

I almost hate to share this information, because I have so much fun predicting when children will awaken. But I must confess, I cannot see the future. I can, however, predict growth spurts, holidays, and the aftermath of troubling events that may create tension and cause your child to awaken in the night.

At nine-thirty p.m., my telephone rang. It was Holly, a close friend whose husband had been recently deployed, leaving her to parent three young children alone. She was distressed. We had worked very hard over the previous months to help her toddler sleep. Our efforts—mostly Holly's—had been rewarded. Mathew had been sleeping eleven uninterrupted hours a night. Suddenly, however, he was waking up and screaming, and Holly was exhausted. "What should I do?" she asked.

"Tell me what's been going on."

"I've been sick, and he's been with a new sitter who came to help out."

"When does he turn fifteen months?"

"In two weeks."

I had a hunch about what we were dealing with. Fifteen months is a developmental milestone. Add a little stress, and you can predict night awakenings. "He's in a state of tense tired," I explained, "over-aroused and unable to sleep. Keep the room dark and massage him. Add a few *shhh, shhh, shhh*s and let him know it's all right. He can sleep."

"Really?" she replied. "Do you think that's it? It's not some horrible new habit?"

"Give it a try. It may take a while to calm him, since he's already distressed, but if you help him now, I suspect he'll sleep better the rest of the night."

"I can do that," she responded. "It helps to just know there may be a reason." The next day, she called to report. "It took forty-five minutes to calm him, but he went back to sleep and stayed asleep for the rest of the night."

A week later, Mathew was noticeably taller and needed clothing one size larger.

Growth spurts affect everyone, infants to adults. If your child is an infant or toddler, mark six weeks, then three, six, nine, twelve, fifteen, eighteen, twenty-one, twenty-four, and thirty months. Growth spurts occur more frequently during this period of time. Learning to roll over, grasp with the finger and thumb, walk, and talk are all developmental surges that have been documented to correlate with disrupted sleep.

If your child is three or older, get out your calendar and mark his birthday and his half birthday. Run a red warning line across the three weeks before it and three weeks after it. This is a heads-up that you can expect more frequent awakenings AND more tantrums, as well as the need for more help doing things you know your child has been perfectly capable of doing.

Fortunately, adults only go through growth spurts about every seven years. It does get a bit easier!

Note holidays

Now go back through and mark two to three days before and after major holidays, especially those where you might be traveling, visiting with relatives, or feeling particularly "stressed" by the event. The anticipation, the disruption of your normal routine, and the increased demands on your time can increase tense energy and lead to more frequent awakenings.

Notice troubling events

You're not finished yet. Think back. What significant troubling events have occurred? Has there been a move or a new baby? Was someone important to your family hospitalized or ill? Has there been a military deployment or a return after a lengthy absence? Was there a major natural event, such as a fire, a storm, terrorist attack, or other significant violent event? Fast forward from the date of the event and note on your calendar the six- and twelve-month anniversary of it. Often we make it through these events only to crash and experience disrupted sleep later. This culprit is often missed because of the time lapse, yet it is very predictable.

Events that are presently occurring also have an impact. The images were bold and harsh. Bodies bloated and stiffened into freakish poses. Victims of nerve gas, the announcer said. Children lay next to fallen parents and grandparents. This weapon knew no mercy. And so, that night, after facing the vivid images on the television, nine-year-old Samantha found herself wide awake at three o'clock a.m., the corpses lying next to her on her pillow. She shuddered, trying to shake the images from her mind, but it did no good. Even in the dark of the night, she could feel them pressed up against her, breathing heavily, white, frothing saliva foaming at their mouths. Sleep, sweet, calm, deep sleep eluded her, thanks to the television news.

This child, who had been going to bed on her own for years, was suddenly insistent that her parents not leave her alone upstairs as she fell asleep. Her parents called me for help. It was the "sudden" change in behavior that led me to ask if there had been any upsetting events lately. Fortunately, Samantha kept a worry journal and was willing to share it with us. It was there we discovered the "corpses in the night."

While you can't predict what will be on the six o'clock news, if your child is seven to ten years of age, make an extra effort to be aware of any violent news stories or major events. Children this age suddenly pay attention to these stories and find themselves awakening in the night.

Finally, mark the days after an adult-only vacation, a conference you attended, or any other separation from your children.

As you do this, you will see that there are a myriad of events with the potential of disrupting your child's, and thus your, sleep. Don't panic. If the awakenings don't occur, celebrate that you skated through them. If they do, you won't be surprised and can plan accordingly.

Minimize commitments

Note the "warning" dates on your calendar and then consider any additional plans you may have made. Expect that near the dates you've marked on your calendar it will be unlikely for you to get a full night's sleep. Since you can predict the potential disruptions, you may choose to cancel a few events and instead take the hot bath that relaxes you. Allow yourself the time to go to bed a little earlier, sleep a little later, or take a nap, knowing that you may need it. Decide ahead of time with your partner who will respond before two o'clock a.m. and who will do so after, so that you are not arguing in the middle of the night. Then, if your child does awaken, you'll have the patience and endurance to work with him. Calmer parents have calmer children. Calm children wake less frequently.

"Daniel must have read those developmental charts before he was born," Marc laughed. "He is so predictable. Now we plan for it and expect that when he hits one, that some time during the night he'll be up. We do our best to slow things down and minimize his stress. We comfort him when he awakens, understanding that it's tension. He goes back to sleep, and so do we. It doesn't turn into a huge battle."

Avoid doubling up tension-producing events

If you're planning a family vacation and have a choice of taking it within a couple weeks of your child's birthday or waiting a few weeks, consider waiting. You'll avoid doubling up a growth spurt and a significant change in routine, both of which have the potential of creating more tense energy. This will make it easier for your child to slip past the developmental milestone without his sleep becoming totally disrupted. Planning also allows you to feel more in control, and reduces your frustration.

Expect to feed your child

Growth spurts are just that, growth spurts, and while your child may not typically need food at night, during a growth spurt, his appetite may suddenly rage. Rather than fighting at three o'clock a.m., consider giving him something to drink. Better yet, give him a more substantial bedtime snack, to prevent the awakening in the first place. This does not have to be the start of a bad habit. A nursing baby nurses more frequently to build up the milk supply to match his growing needs. Once it's built up, the frequency of nursing drops off again.

Once your child is taking in solid foods, the extra snack during growth spurts can be temporary. You can let him know that you will provide it, but soon—within the next few weeks—he won't need it anymore, and it will stop. I suspect your child will surprise you with his willingness to "give it up" when his needs have been met and he's through the growth spurt.

If he's reluctant, observe closely for the change in his physical size, or his cognitive, social, and emotional skills. You'll also notice less clinging or demands to be with you. When you see it, you will know that it's now all right for you to start the gentle nudge.

Begin practicing at the first awakening of the night, especially if it is before three o'clock a.m. He'll be more tired then, and the pressure in his body to return to sleep stronger, thus increasing the odds for success. Instead of offering food, simply give him a sip of water. Expect that you may have to get up to comfort him. If he's young, you may need to hold him, not in a feeding position but over your shoulder, swaying slightly, offering comfort but not food. You'll realize that while he may protest, his cries are half-hearted. He doesn't like it, but he can do it, and, as a result, everyone will be able to get more sleep again.

If he awakens a second time in the same night, you may choose to feed him this time. While this may seem confusing, you are breaking the process down into tiny steps. After three o'clock a.m., his sleep debt is less, reducing the pressure for him to go back to sleep, and the likelihood that he's actually hungry is greater. I strongly suspect there will be another opportunity to practice again the following night, shortly after midnight or even before, when, as earlier, it will be easier for him to practice returning to sleep without eating. Once the early-night awakening has disappeared, you can tackle the later one.

If in the process, however, you recognize that your child's protests are vehement and the screams heart-wrenching, he's letting you know that he's not quite through the spurt and still needs a little more time. Try again in another week or two, unless he's temperamentally very intense and always screams loudly. Then you have to go with your "gut," recognizing that while he may not appreciate it, he is ready to work through it.

Include more soothing, calming activities

Tension creates arousal. When you realize that your child is in a state of tense energy or tense tired, add more soothing, calming activities to his day and to his nighttime routine. This is especially true if you have recently encouraged him to give up a favorite "lovie." It was a tension buster for him, and, as a result, he needs a replacement.

Instead of ten minutes of massage, plan for twenty or even thirty. Stroking stimulates the production of chemicals that inhibit the stress hormones. Add a second storytime during the day. Allow more time in the bath. Make an extra effort to sit down for family meals, and include chewy foods. Encourage exercise and/or listening to music. All of these things reduce tension and make it easier for him to fall into a deep, restorative sleep, from which he is much less likely to awaken.

Expect, too, that he may need more support as he "switches" to sleep. If he goes down crying, it's more likely that he will awaken later. So, support him and help him, knowing that it is temporary. If necessary, stay with him or nurse him, but again, as he falls asleep, if at all possible, attempt to move away slightly and remove the nipple, so that his final "switch" to sleep is independent. This is true even if you share a family bed.

Let your child know you will respond

Knowing that you will respond, even in the middle of the night, gives your child a gift of security that lasts a lifetime. Do NOT leave your child screaming in the night. You wouldn't do it during the day, and you don't have to do it at night, either. We really do not know the long-term emotional costs of this strategy, nor do we know which children are most vulnerable to anxiety as a result of it.

How you respond to your child will vary. Early in the night, the urge to sleep is strongest, and it's a good time to practice returning to sleep independently. When your child wakes up, give him a few minutes to return to sleep independently before you intervene. Remember that he's going through a normal sleep cycle, and may not be fully awake. If, however, your child is very intense, even at this time of the night, it may be important to catch him quickly. Doing so may mean the difference between a two-minute arousal and a two-hour one.

Start simply with the least intervention possible. A statement like "We're here, it's all right, you can go back to sleep" may be enough to reassure your child and send him back to dreamland. A pat or a few strokes of his brow may do the trick. As the night wears on, it's more likely that it will take more intervention to get him back to sleep. This is where the feeding may come into play, holding him or tucking him into bed with you. Your child isn't being stubborn at four o'clock a.m., it's simply that his sleep debt is lower and it's harder for him to return to sleep.

Gather me in

Seeking comfort from those we love is a normal response to distress. When your child is in a state of tense energy and tense tired, he's more likely to seek contact with you, even in the middle of the night.

When children are distressed, you cannot threaten, force, or beg them out of it. You have to address the feeling. If you are comfortable doing so, let them in your bed. Enjoy the cuddles and go back to sleep. If you prefer to sleep next to your partner, and aren't worried about your child rolling out of bed, let your child know that she can come into your bed, but she must sleep either next to one adult or the other, but not in between. And if you can sleep with a child in your bed but your partner can't, consider Christine's technique: "The kids know, there's a 'yes' side of the bed and a 'no' side. They very carefully tiptoe past their dad and slide in next to me."

Or you could try putting your child to bed in his own bed, spending time with your partner and then, if your child needs to be "gathered in," moving to his bed. Just be sure to get snuggle time with your partner. That's what Franklyn did.

Inevitably, two-year-old Molly had been awakening while her parents were in deep sleep. It was very disruptive to them until they came up with a plan. "We put Molly to sleep in her own bed," Franklyn told me. "Then we have private time. When we are ready to go to bed, we pick her up and put her in bed with us. She still seems to be waking, but now, because she's in our bed, she simply rolls over and goes back to sleep, and we're not disturbed."

If you aren't comfortable with your child in your bed, consider letting him sleep near you, either on a cot or in a sleeping bag on the floor. That's what Megan did.

Five-year-old Brandon had been appearing in his parents' bedroom every night several times a night for weeks. "We'd march him back to his own bed every time," Megan said, "but inevitably, within an hour, he'd be back again. One night I took him back ten times."

Since they could not sleep with him tucked into their bed, I suggested purchasing a sleeping bag and putting it on the floor next to them. "We bought him a bag," Megan told the group two weeks later. "Before bedtime, we would lay it next to our bed and tell him that if he awoke in the night and needed to come to us, he could crawl into the sleeping bag and sleep near us. The first three nights, he came in around two o'clock a.m. and slept until eight o'clock the next morning. However, last night he slept all night in his room without waking once. A big WOW!"

Simply knowing and trusting that you will respond and "gather him in" when needed is frequently all your child needs to reduce his tension and allow him to sleep more soundly.

If you don't have a bag, keep your child's sheets and blankets simple, so that he can just bring them in with him. That's what Ginger did. "I made duvet covers that could be laundered easily and just used a bottom sheet. The kids bring their duvet with them and either make a nest on the floor or crawl up into the big chair."

If your child insists that the sleeping bag is not "good enough," work together with your partner. "We decided on the strategy ahead of time," Paula informed the group. "We recognized that we were meeting her need to be near, but that we also needed our sleep. She's a little wrestler at night, and neither one of us can sleep if we let her in our bed. She's four and a half, so we explained the plan to her. The first night, she came in and protested. She was cold. So I got up and found an extra quilt to put over the bag. She crawled in, and I zipped her up tight. Five minutes later, she wanted into our bed. That's when my husband rolled over and said, 'You heard your mother. The rule is, you can be close, but we all need our sleep.' She huffed, but then she rolled over and went to sleep." When two adults say "this is the rule," children do accept it. The "backup" adult helps your child to know that she can come for support, but that "no" is "no."

If you are not comfortable tucking your child in bed with you or bringing him into your room, there are other ways to provide sensitive care and still let everyone sleep.

Put siblings together

In most of the world, siblings sleep with one another. Children are not expected to sleep alone. If your child likes to cuddle, but you can't sleep if he cuddles with you, consider letting your children sleep together.

"I was eleven before I ever slept alone," David remarked in class. "I don't remember getting up. I just moved closer to my brother."

Often, the concern with siblings sleeping together is their keeping one another awake. Kylie's mom solved this problem. "When I was a child, we were three girls, and all of us slept in the same bed. Fifteen minutes before we settled down, we had 'giggle time.' My mom expected it. I never remember it as a problem. But the rule was that if one fell asleep, you couldn't wake her up."

If "giggle time" would extend for hours in your home, consider putting the children to sleep in different rooms, then moving the one who is most flexible and the soundest sleeper. Or simply tell them that if they awaken, to go snuggle with a sibling. The next morning, it may be interesting to walk around discovering where everyone ended up. Most important, they will have slept. There's something very comforting about listening to the breaths of another person in the night. Adults often tell me they are closest to the sibling they slept with.

When you meet the needs of your children, while at the same time gently structuring so that everyone in the family can get more sleep, you'll find that they will sleep better. You might worry that you're being too "soft," but sensitive care blocks the stress reaction. The more you can reduce the stress reaction, the better your children will sleep. It's also most effective if you include them in the plan.

Work together with your child

Clarify your vision

Every morning at three o'clock a.m., Micah arrived at his mother's bedside, ready to climb in. She didn't mind letting him, except that he kicked and often interrupted her sleep for the rest of the night. When we talked in class, she realized that she'd been so accommodating to Micah's needs that she'd never told him that he was supposed to sleep until morning. Children do not necessarily know that they are expected to sleep all night long, until you teach them. Be sure that your child has a clear vision of where you are going. When you put him to bed, you may include in your nighttime ritual a little ditty, such as "sleep tight till the morning light." And be sure to reinforce success. In the morning, you san say to him, "What a good sleeper you were. You slept all night long!" Or "You did it. You stayed in bed all night long." Or "You slept without waking up Mom or Dad. You helped everyone get enough sleep."

If your child awakens and needs help going to the bathroom, practice with him during the daytime, when everyone has more patience. Walk him down the hall from his room to the bathroom; let him know that "Big people go to the bathroom all by themselves in the night. They don't wake up others. Pretty soon you'll be ready to do it, too." Then, each night, ask him, "Is this the night you're going to surprise us and take yourself to the bathroom?" One day he will.

Teaching coping strategies Waking up and calling for Mom and Dad is one strategy for coping with arousals in the night. What many children don't realize is that it's not the only one. Brenda explained: "Now I tell him, 'You can always come to Mom and Dad, but first try adjusting your covers, fluffing your pillow, taking a sip of your water, or cuddling with your bear. If that doesn't work, then you can come to Mom and Dad.' " When your child calls for you, you can ask him: "Did you try adjusting your covers or getting a drink?" If he didn't, you can remind him that next time, he can try that first. A sticker chart with a prize after two or three "successful nights" may help to teach the new skill.

TIME—MAKING SURE THE BODY CLOCK IS "SET"

Sometimes, as you work with your child, you realize that the issue isn't tension. There are no definable growth spurts, no troubling events have occurred, and there haven't been any significant separations or stressors. It's time to check your schedule. Have you inadvertently been giving your child jet lag, which leads to more frequent awakenings?

Review your schedule

Stop and think a moment about your schedule. Has your child's bedtime been delayed? Did you skip or shorten a nap? Has he been waking in the morning at different times? What about meals? Have they been erratic? Have you recently traveled across a time zone? All of these decisions can disrupt the clock and lead to more frequent awakenings. That's what Jenna discovered.

"I've come to realize that no matter what time he goes to bed, he gets up at the same time. So, if our scheduling is bouncing around, he's always losing sleep. In the past, I kept thinking that if he went to bed later, he would get up later, but it never happened that way. What he did start doing was waking up in the middle of the night. For the longest time, I saw it but I didn't know that was why he was waking."

Go back and review your ideal schedule. Make adjustments, if needed, but get your family back on a predictable routine. It will make it easier to get the rest you need.

Avoid skipping or dropping naps too early

It's very common to hear from an exhausted parent that their normally "good sleeper" is suddenly not sleeping. Often, after talking, I learn that they have recently made the decision to "drop" their child's nap or shorten it, because the child was refusing to go to bed until ten o'clock or eleven o'clock p.m. Overtired children do NOT sleep well. They wake more frequently in the night and even tend to wake earlier in the morning.

In contrast, well-rested children are less tense, and their body clock is set, and, as a result, they sleep more and better. If your child is having trouble going to bed on nap days, check your overall schedule. Has it been unpredictable? Often, the refusal to go to bed is linked to "jet lag" rather than having an unnecessary nap. *Consistently* maintaining a siesta or naptime will increase the odds that your child will sleep all night long and still go to bed at a reasonable hour.

Check light exposure

Has the weather been bad, so your child hasn't gotten outside to play? Have you been busy and less likely to monitor your child's television viewing or computer-game time, so it's gradually increased? Limited exposure to morning light, or exposure to light at the wrong time, especially in the evening, can lead to more frequent awakenings. "I hate to admit it," Jennifer told me, "but after the holidays, we really fell out of our routine. It was cold and dark, and the kids got the flu. It was so much easier to just let them watch television or play a computer game, but I finally realized it was really costing all of us more disrupted sleep. It makes such a difference when we stick to our routine and turn off the electronics in the evening."

If your child is waking frequently, turn off the television and get outside, especially in the morning, and it's very likely that you'll stop having visitors in the night.

Check your bedtime routine

I awoke in the dim light, confused and startled, not knowing where I was. The light I'd left on in the bathroom crept into the room and gave me a chance to orient myself. I was in a hotel room in Boston, alone. The warmth and pressure of my husband's body wasn't there. The sounds in the hallway were foreign to me. I heard voices and a car door outside. Once again, I startled and was then wide awake. Am I safe, I wondered? I lay back, breathing deeply, trying to quiet the thoughts in my mind, and waiting for sleep to come again.

Regular REM sleep periods serve a survival purpose. If the sensory input from the environment is different from what you are used to, or from that registered as you fell asleep, an alarm function wakes you up. Researcher Avi Sadeh found that when a child falls asleep under the same conditions as those that prevail during the night, he is less likely to wake up during light-sleep phases. If a child registers the presence of a parent or a pacifier in his mouth when falling asleep, he is easily woken up during light sleep when unconsciously registering that the parent or the pacifier is no longer present.

If your child is awakening during the night, especially at predictable times, it may be that his brain is noting a change in the environment. So, check, did you turn off the hall light or the fan AFTER he went to sleep? If you did, consider either leaving it on all night, or turning it off just BEFORE he falls asleep. The more consistent the conditions in the environment are from the time your child falls asleep and through the night, the easier for him to move from one sleep cycle to another without awakening. Now, as I state this, let me also say, know your child. That's what Patti discovered. "If I turn off Michael's fan, he'll arouse in the night. But if I DON'T turn off the nightlight next to Leah's bed, she wakes." Observe closely what cues in the environment seem to be most helpful for your child to feel safe enough to sleep.

YOUR CHILD'S TEMPERAMENT MAY LEAD TO MORE FREQUENT AROUSALS

Reading through the tips for helping your child sleep through the night, I suspect there have been moments when you realized that while a recommended strategy may work for many children, it's not going to work for yours. That's important to know, and you have permission to tweak the plan to fit your child.

When your child is intense

Intense children wake more frequently in the night; it's a documented fact. They're biologically wired to arouse more easily, and, as a result, are much more sensitive to tension and disruptions of their body clock. This child really needs you to help him manage the tension during the day, including lots of soothing and calming activities and keeping his body clock set with a regular schedule. Preventing those arousals in the first place helps him to manage his tension more effectively and stops the middle-of-the-night screams. If he does wake, know that you need to respond quickly, so that he doesn't go into full arousal.

Respond quickly

Contrary to ALL the advice you'll receive about not rushing too quickly to respond, the intense child needs you to catch him before he goes over the edge. "That's so true!" Carrie gasped. "Once Damon is upset, it's unbelievably difficult to get him settled again."

While some people may advise you that to do so reinforces a negative reaction, what we now know is that it doesn't. Megan Gunnar at the University of Minnesota tells us, "When it comes to brain cells, what fires together wires together." The intense child will automatically "fire" brain cells to "alert." What he doesn't automatically "fire" are brain cells to calm, but when you step in and help him quiet, the two begin to match up and ultimately "wire together." That's why the intense adult who has learned to manage his intensity well reports that he feels hot, but not explosive. His brain, we believe, has developed a new way of functioning.

Listen very carefully to your child. Learn to discern the differences in the tone of his protests. Sometimes, you'll hear, "I don't like this, and it's hard, but I'm doing all right." When you do, you can pause for a moment before responding, and let him practice. But if you hear a wail that clearly communicates "I'm overwhelmed, I need help," it's time to step in. By noting the differences, you can bring him back from the brink and save both of you a great deal of turmoil.

If you are intense, too, make managing your own tension a high priority in your day. When you respond to your child, take a deep breath. Remind yourself he needs your help. It's all right. All children awaken. You can help him, and soon he'll go back to sleep. The payoff for those nighttime rendezvous may come years later, when your teen or young adult chooses to flop on the end of your bed and regale you with stories of his day.

If your child is sensitive

Everyone unconsciously registers the sensory impressions of the environment, but it's the sensitive child who is especially aware. Before you start baking or preparing the next day's meal while this child is sleeping, decide whether or not it will disrupt his sleep. It's the sensitive child who awakens to the smell of something burning, or especially fragrant. Smells will linger in the night, and it's often easy to miss that the sauce you cooked at eight o'clock p.m. is now alerting your child at two o'clock a.m., but it can.

The sensitive child also awakens to the slightest noise: water dripping off the roof on a spring night, the creak of the floorboard, the closing of a door. Keeping a sound machine or fan running in his room can help him sleep through the night.

Carefully monitor the stimulation during the day. Overstimulation keeps this child alert at night. Sue was delighted to inform me that her highly sensitive son William had slept through the night. "It took four weeks of carefully monitoring the stimulation during the day," she told me. "I reduced the number of errands and turned off the television as background noise. Initially, I thought it wasn't working, but I guess he just needed some time to wind down. Now he doesn't come to full arousal anymore, and instead goes back to sleep."

If you are sensitive, be aware that you may alert to every breath of your child. Give yourself permission to wait to be certain that he does need your assistance. Allow yourself to sleep where it is quiet.

If your child is slow to adapt

Transitions are tough for the slow-to-adapt child, and this is also true as he moves from one sleep state to another. Keep his bedding and "lovies" simple. When his blanket or bear falls off the bed, or the pacifier disappears, it may disrupt his sleep.

Know, too, that once your slow-to-adapt child is awoken, it will take longer for him to return to sleep. Again, this is not because he's trying to be difficult, but simply that his body struggles more to shift from one stage to the next. You can help him by patting him and then gradually reducing the amount of time you do so. Stay near, don't abandon him. Know that he's working hard and will get there, but it does take time and practice.

If you are a quick-to-adapt parent, recognize that you may be so good at adapting to your child's needs that you have not begun to nudge him. It's important to catch yourself and practice with him falling asleep and returning to sleep on his own, before you finally do hit your point of exhaustion.

If you are slow to adapt yourself, do your best to get to bed early enough that you can get four hours of sleep before your child awakens you. Otherwise, it's so hard for you to shift that you may wait until your child is too upset to return to sleep without a great deal of assistance.

If your child is high-energy

The high-energy child doesn't magically stop moving during sleep. She needs a sleeping space large enough and soft enough that when she throws out an arm, or kicks, she doesn't hit a hard object that might startle and awaken her. If you have the choice, this is the child who needs the double- or queen-sized bed, and while the molded plastic race-car bed may be cute, the highly active child is likely to bounce off the hard sides and awaken.

MEDICAL ISSUES CAN CAUSE AWAKENINGS

Sometimes, as you work with your child, you realize that tension, time, and temperament are not the culprits that are leading to frequent arousals. If this is the case, it's time to consider medical issues.

Infections

B. Eckerberg Leksand from Falum Hospital in Sweden found infections to be a common reason why children awaken during the night.

When your child is uncomfortable, fussing and fuming during the night, and your attempts to soothe and calm have been ineffective, your child may be coping with an infection. Remember, it's not that he won't stop crying, but that he can't. It hurts. Wendy was delighted to report in class that she had recognized her son had an ear infection even before her doctor could see it, because of the change in his sleep.

Snoring

Sometimes you may not even be aware that your child is awakening, but if he snores, it may be an indication that he is suffering from sleep-disordered breathing in which he appears to be sleeping but is actually alerting frequently during the night. One in nine children, boys more likely than girls, has been found to have sleep-disordered breathing. Excessive weight is also associated with the condition.

Allergies, asthma, and other issues

Frequent night awakenings can be an indication of other medical problems, too, including asthma, digestive issues, reflux, constipation, allergies, and optic nerve damage. In any of these cases, contact your doctor for full testing.

Night terrors—what it is and what you can do

One of the most disconcerting awakenings in the night is a sleep terror. "I heard the scream and dashed into his room immediately," Diane told me. "He was sitting up, crying and thrashing and shaking. His eyes were wide open, so it looked like he was awake, but when I spoke to him, he didn't respond at all. It really frightened me. I had no idea what was happening."

During a night terror, your child appears to be awake, but he's not. He may talk or babble, sit up in bed, or even walk around the room. He seems terribly frightened, but does not find your touch reassuring. Instead, he may push you away.

Sleep terrors are characterized by a sudden episode of intense terror accompanied by screaming or crying during deep sleep. They usually occur in the early part of the night, often within two hours of your child falling asleep. Occasionally, they'll also strike during naptime. They occur most frequently in children ages three to eight years, and usually disappear by age ten. Your child may experience one and then no others, or he may have a couple in one night, or one each night for a week. Families often report a tendency for them to "run in the family."

It's not certain what triggers sleep terrors, but there seems to be a link with being overtired, stressed, or anxious, having a fever or taking medications.

So, how do you respond? Your first task is to keep your child safe. Go to him. Avoid calling his name or asking questions about what's wrong. Even though his eyes are open, he's asleep. Reassure him. Let him know you are there, that he can go back to sleep and that everything is fine. If he'll let you, lay him back down. If he doesn't want you to touch him, don't, but stay near to ensure his safety.

Then put on your detective's hat and look for the event or issue that may be triggering this. When parents were asked when their children experienced night terrors, their answers varied.

"Tatum started having them the first night I went back to work."

"John's started when school began."

"Trisha woke up screaming after a frightening ride on a ferry."

"Tommy was up four times a night, after it was announced that his teacher was leaving."

"A jam-packed day seems to set Ben off."

And Marc wakes up screaming when there have been lots of tantrums during the day.

Once you've identified the culprit, work with your child to allay her fears. You can write a letter to the teacher who is leaving, and reassure your child that she can stay in touch. Let him know where you'll be and when he can expect to be with you again. Ask what will make it better and then do the things that need to be done. Put him to bed earlier so that he's not overtired. Ease the tension, and your child will more easily make it through the shift from deep sleep to light sleep without screaming.

Nightmares

Nightmares are scary, long, vivid dreams that awaken the sleeper in the later part of the night, usually after midnight. Unlike night terrors, nightmares can be recalled immediately afterwards, or even in the morning. Your child will also respond to your nurturing touch. Nightmares occur more frequently in children age three to six and tend to decrease over time. Like night terrors, they are often the result of a recent stressful event or change. A nightmare can also indicate an illness with fever.

When your child experiences a nightmare, reassure him. Stay near to soothe and calm. Avoid television, scary stories, and frightening games before bedtime, and high stimulation during the day. Older children may need as much support and help as younger ones.

If your child is plagued by nightmares, talk about them and rewrite the story. If a monster is chasing him, reverse the story line so that he turns and chases the monster away. If he's being held down, tell him how he uses superpower to free himself.

When it's been a tough night

When, despite your best efforts, it's been a rough night, recognize that energy and patience levels are low. Review your plans for the day. Eliminate what you can. It's much easier to be tired at home and in private than in public, surrounded by crowds of people. Reduce the demands on your child. Expect him to require help getting dressed or organized for the day. Odds are he'll need extra support with homework. Plan, too, for both an early nap and bedtime. And don't forget yourself. Nap if you can. Call a friend to give you a break, and take advantage of restaurant delivery.

Celebrate successes

Being a parent is a very tough job, made even more challenging when you are awakened in the night. So, each morning, stop and ask yourself: was I a kind coach during the night? Did I gently respond to my child's needs AND patiently provide structure for him, tenderly teaching him how to sleep in our family? When you see yourself as your child's coach, you realize that he has only been on earth a very short time. He needs to sleep, wants to sleep, but isn't quite sure how it's supposed to work in your family.

Be clear about your goal. Expect it to take time and repetition. Meet your child's needs yet gently nudge her in the direction of working with you and respecting the needs of the whole family. Do it in a way that keeps her in a state of calm. It is not a quick fix but a lifelong process of learning that her needs are important and she deserves to have them met. At the same time, she is part of a larger family unit that works together. Initially, you will have to do all the giving, but gradually, your child will begin to reciprocate, even in the middle of the night. And one day, you really will forget how difficult it was.

FOURTEEN

NAPS AND SIESTA TIME

Eliminating the late-afternoon meltdowns

"I tried to create a nap schedule for him by saying we go to music and then we nap. Now I realize that he needed a nap at 9:30 a.m., but we were in music class, so he fell apart."

—Ada, the mother of Mathew, 9 months

When I was a child, my grandparents had a cottage upon the bluff, sixty-three steps above the shores of Lake Pepin. It was here that my cousins, sisters, and I spent much of our summers.

Painted bluebird-blue, like my grandfather's eyes, it stood on cement blocks sheltered under towering pines. A screen porch wrapped around the front and one side, housing a round oak table for card games and summer meals, two daybeds, two single hospital beds that mysteriously raised your head or your feet, and a row of unmatched wooden rocking chairs.

Inside the cottage, a potbellied wood stove dominated the center room, where another round table sat. Off the center room lay two small bedrooms, separated by a partition only six feet tall. My grandfather and father could look over it.

Hot water for bathing or dishes came from the tea kettle on the stove. The only luxury was a flush toilet.

My grandmother, who stood four feet ten inches tall and was normally quiet and gentle in nature, had one ironclad rule—siesta after lunch.

Adults and the very young ones often napped during our enforced siesta, while the older kids were left alone to draw or read. If we were lucky, we would discover Grandpa's stash of *True Detective,* and, after fighting over who got the high-backed rocking chair, buried ourselves in the dramas that unfolded. No one, not even the teens, snuck out.

Looking back, what's interesting to me today is that no one questioned this ritual. The rule stood for everyone, and while some might have lamented the sun they were missing, they still took a break. Perhaps it is also this ritual that allowed as many as fifteen of us to not only live together but to delight in life at the cottage. Perhaps, intuitively, my grandmother knew what researchers have now documented: naps are good for everyone.

A different point of view

Unfortunately, naps have gotten a bad rap. We seem to find them a crutch, something for babies, an inconvenience to get rid of as soon as possible. Full-day pre-kindergarten programs often drop them from the schedule as a waste of time, a luxury that four-year-olds can no longer afford. And, in our busy lives, they're easy to lose in a rush of errands and activities. That's why one of the most common questions I am asked is: "When should my child stop napping?" Researchers today are discovering that the answer is NEVER. The challenge, of course, is how, in a world focused on production and achievement, do you find the guts to stand up and insist: "Naptime!"

Why naps are essential We're not quite certain what happens while we're napping, but Dr. Mark Mahowald, director of the Minnesota Regional Sleep Disorders Center, states, "It's possible that after a morning of hard work, the brain's inbox piles up too high. When that happens, fatigue may signal that the brain needs time 'offline' to process everything."

Sleep scientists know that, during a good night's sleep, newly learned information and skills are consolidated. Now they are discovering that naps also allow that consolidation to occur.

Dozens of studies show that napping increases alertness, productivity, and mood. A study completed at Harvard University found that individuals who napped were better at retaining information and concentrating. Naps improve motor skills, including reaction time, and are crucial for healthy growth and immune function. In the rush of a busy day, your child's nap helps to restore self-control. Without it, all energy has to go into staying awake, which also leads to more bedtime battles and middle-of-the-night awakenings because your child is overtired. And if that's not enough, research also demonstrates that a child who has been awake for eight consecutive hours is four times likelier to have accidents.

The fact is that naps are NOT a luxury, but are ESSENTIAL for growth and development, health, mood, performance, and satisfying relationships. If you need an excuse to explain to your friends why you're not signing up your child for an activity, or leaving an event so that your child can get his nap, let them know that brilliant people, including Thomas Edison, always took their naps. Or, as one of my students once quipped, "Now, when we leave our friend's house so the kids can nap, we say, 'We're heading home for cognitive consolidation time.' No one wants to argue with that!"

Note your child's behavior on no-nap days If you're still not certain that your child needs a nap, go back and review the sleep chart you completed in chapter 8. On the days that your child's nap was skipped or abbreviated, note what happened to his behavior and the total amount of sleep he got in that twenty-four hours. In class, when we took a look at the charts, we discovered that two-year-old Jennifer was a different child on her no-nap days. When her nap was skipped or shortened, she didn't make up the sleep at night. Instead, she slept only nine hours in a twenty-four-hour period. On those days, she was a monster. But on the days that were less stimulating and hectic, she slept a full two hours in the afternoon and another ten at night. On those days, Jennifer was a model child.

Four-year-old Emily seemed to thrive on activity and often complained she was bored if there was a lull. She went to dance classes, preschool, played with friends, and shopped with her mother. In terms of activity, one o'clock in the afternoon was no different from nine o'clock a.m. in her day. But the log showed that, by late afternoon, Emily would melt down. The intensity of the tantrums grew worse as the week wore on. By Friday, she was a bear. Stop and watch carefully. In all likelihood, your child's behavior will show you that she really does need her nap, even if she initially resists it or appears to have given it up long ago.

What naps look like Naptime in most cases means sleep time, but it can also be a siesta, during which you or your child take a break. During this time, your child may read, draw, listen to music, do puzzles, or simply lie down and daydream. It doesn't mean watching a video or favorite television show. At the Cambridge Friends School, children are asked to merely sit quietly. School head Mary Newmann states, "Beginning in infancy, children are bombarded with noise, stimulation, and instant gratification. Quiet time; it's virtually programmed into children never to have it. But if you want to raise children who can think critically, who can solve problems of all kinds, they need the chance to think uninterrupted."

After forty-five minutes to an hour, if your child has not fallen asleep, siesta time is finished. At Paidea Child Development Center, one-third of the kindergarten students nap during siesta, and in the four-year-old program, over three-quarters do.

Even school-age children can benefit greatly from a break in their day. "Our lives changed when we added the quiet hour in the afternoon after school," Susan told me. "The kids can nap, read, draw, or write. It's so great for everyone, including me. When they resist it, I tell them, they don't have to sleep but we all need a break. It gets us through all of the other activities and dinner so much more easily."

The length of your child's nap will vary. Infants and toddlers may sleep up to three or more hours in a stretch. Most preschoolers nap one and a half to two hours. If your child is napping more than three hours, and you are experiencing difficulty getting him to sleep at night, wake him after three hours. You want the longest periods of sleep consolidated at night, not during the daytime, but if your child is napping for long stretches AND sleeping well at night, let him sleep!

You and your older child may find that a mere twenty or thirty minutes is all that's needed to give you the energy for a whole "other day." Winston Churchill was convinced of the efficiency of naps and is quoted as saying, "Don't think you'll be doing less work because you sleep during the day. You will be able to accomplish more. You get two days in one—well, at least one and a half."

Where to nap If, at this moment, you are thinking, I've tried quiet time in my child's room and it just became a power struggle—don't worry. Where your child naps and where she sleeps at night do not have to be the same. Your child may nap in her usual sleeping space, or she may prefer a different one, like the couch, a cot, your bed if she doesn't sleep there normally, or any other comfortable site. While it may not matter for your child where it is, once it's been decided upon, consistency of the site remains important, because it helps to cue that it's time to sleep when you go to it.

A little creativity was the key for Christine. "Even if my son got up at four thirty a.m., I couldn't get him down for a nap," she told me. "In the past, I'd tried to get him into his room, but it became a power struggle, with him yelling, 'I'm not going. You can't make me!' When I did get him to his bed, I'd read to him, but then he'd jump on his bed and freak out if I tried to leave him. Yet I could tell from his behavior that he was really tired.

"One day, after lunch, I simply invited him to lie on a blanket on the floor and read with me. I closed the shades in the main room of the house, got a blanket and pillow—not his, because that started him screaming again, but rather those off the couch—and started reading, and he fell asleep. Now I tell him that after lunch we're going to read. It's usually about one o'clock or a little after. He falls asleep within twenty minutes. And the funny thing is that he's now started asking to go to bed at night instead of begging to play ten more minutes like he did before. I never believed the saying 'sleep begets sleep,' but in our case, it's absolutely true."

If a bit of flexibility on your part will take the "fight" out of naptime, feel free to choose a site that works for you and your child. Some children will need a space that clearly says 'time to stop and rest,' while others will only need a comfy spot. More important than being in bed is stopping to rest.

When your child wants you for nap Four-month-old Joanna naps—but only in her parents' arms. Two-month-old Thomas will nap—if his mother lies down with him. Otherwise, he awakens the moment she lays him down, or sleeps for a mere twenty minutes. Young infants under six months of age, like Joanna and Thomas, who prefer napping with you, are often viewed as "problem nappers" whose style needs to be fixed. In reality, napping while being carried has been a fact of life for infants since the beginning of time. And there's actually something very interesting about these babies. If their parents do help them to sleep by holding or staying near them, at night they will sleep in their "nest," even if it's a crib or bassinet. Not only will they allow themselves to be put down (awake or asleep), but, often at a very young age, will sleep uninterrupted five to nine hours through the night. So, what's up?

I call infants like Joanna and Thomas "leader babies," because, if you allow them to, they will lead you to balance in your life. Just as you are in tune with your baby, your baby is in tune with you. There's a dance of hormones that goes on between mother and infant. When you sit down to nurse a baby, stop to rock and watch him, or simply hold him, Mother Nature releases soothing, calming hormones into your system. This "slowing down" is one of the best tension busters available for meeting the demands of life with a young infant. Rather than "spoiling your baby," a recent study reported in *Pediatrics* shows that lots of cuddling can promote baby's development. And parents who cuddle with their little one regularly reported feeling more bonded to their baby and more confident in caring for him.

Emily wasn't aware of "leader babies" when she came to class. On the first day, she had slouched in her chair and lowered her eyes as she confessed in a soft voice: "Sometimes I lie down with Emerald in the afternoons and nurse her to sleep. If I do, she'll sleep for two hours. If I just lay her down in her bassinet, she's up in thirty minutes. I know it's so bad, but sometimes it's just the easiest thing to do." She grimaced in embarrassment.

"Let's talk about why this feels awful," I said.

"Well, it's so decadent, to be lying down with my four-month-old in the middle of the day. There's housework to be done and laundry to fold. She's not soothing herself. She's falling asleep nursing. I know I'm starting a terrible habit. Everyone tells me so. And then my husband comes home and asks, 'What did you do all day?' How do I admit I slept? I should be exercising, or doing something, anything. How can I just lie there and snuggle with my daughter?"

The work ethic remains alive and well in our culture, but sometimes, by letting those hands idle for a bit, we actually enhance their performance later. Emily, by following her natural instinct, hit upon a very important truth. Children sleep better when their parents aren't stressed. By following her "gut," Emily slowed down, soothed and calmed her baby and herself. She wasn't wasting time. She was allowing her baby to get the sleep she needed for growth and development. It was also the sleep Emily needed to be more patient and sensitive to the cues of her daughter AND more open and responsive to her husband.

If you have older children and a "leader baby," you'll have to be creative. In the morning, you may put your baby in a kangaroo carrier and snuggle together with the older kids for reading time. Or you may use the time to sit down at the table, while they draw or paint. Not only will you rest, but also connect, making the addition of another family member easier on the older ones. In the afternoon, make it siesta time for everyone. I suspect you'll be pleasantly surprised by the energy and efficiency you gain, and by how much you'll be able to accomplish as a result of "listening to the cues."

And if you worry about the other kids missing out on an activity because you've chosen to slow down, there're two very important things to remember. First, research has found that time spent in conversation and interaction with parents is more predictive of future success than involvement in one more activity, and second, it's all right to teach our children that sometimes we make sacrifices for the good of the family.

It's not going to be forever. Continue to practice with your "leader baby" by laying her in a crib or a bed at least once a week. Ultimately, she will begin to accept it and merely roll over and go to sleep.

Implementing naps

When I introduce the topic of naps at my seminars, it is met with groans of frustration.

"My child stopped napping the day she turned two."

"I try to put him down for a nap, but he fights it and doesn't fall asleep."

"Sometimes he sleeps and sometimes he doesn't."

"He naps at day care, but absolutely refuses to nap at home."

These are all common reactions from parents. And it's true; the trouble with naps is that your child has to participate in them. But if you anticipate a problem, you can plan for success, much in the same way as you do for bedtime.

Seeing Potential Trouble Spots

Find your child's nap "window"

It has become almost a quaint throwback that sleep is a priority. It's much more common to register for three classes and strive to complete them no matter how exhausted the schedule leaves you and your child. But, just as your child has a "window" for sleep at night, he also has an opportune "window" for naps. Right after lunchtime, there is a natural dip in energy levels for all individuals preschool age or older. That's why you often feel sleepy after lunch—it's not what you ate, but rather a natural dip in energy level. Mother Nature is signaling that it is time for siesta. Frequently, we fight her pull by consuming a caffeinated beverage, a bit of chocolate, or doing something to keep us going.

If we listen to her cues, however, and give her a chance to help us out by planning downtime into our routine, Mother Nature will gently close our eyes and send us off to sleep. So, if your child is a preschooler, shortly after lunch, start noting signs of slowing down, going for a "lovie," becoming a bit cranky, craving something to "charge up," or losing coordination. Often, these cues are very subtle and short-lived, so, watch carefully. You are aiming for this "window." If you establish a siesta right after lunch, your child will have time for her nap and still be awake by three o'clock p.m., which means that she'll be ready for bedtime early in the evening. If naptime runs later, expect bedtime to be later as well.

A typical schedule for a preschooler might be to arise at seven o'clock a.m., have lunch at noon, nap from one o'clock to two-thirty, followed by sleep time at eight-thirty p.m. What you'll find with preschoolers is that there is generally a five- to six-hour break between sleep periods. You can use this as a guide to help you find your child's "window." Now, if your family prefers a later bedtime and your child has the opportunity and will sleep in the late morning, then naptime can be moved accordingly. If, however, your child wakes early, no matter what time he goes to bed at night, it's important to honor his body's schedule and put him to bed early, when he hits his natural "window" for sleep. Then you can plan to spend time with a well-rested child in the morning, rather than a crabby, overtired one at night.

Recognize common culprits behind nap resistance

The national recommendation for preschoolers is eleven to twelve hours of sleep in a twenty-four-hour period. In my practice, I've found that once they are given the opportunity to nap, the reduction in tension caused by sleep deprivation frequently leads the children to sleep twelve and a half to thirteen hours. If your child naps but then is up later than you'd like him to be at night, there are some other options you can explore rather than cut his much-needed nap.

First, determine how long you have been encouraging naps. It takes two to three weeks for the body clock to move. If you've just reintroduced naps, hang in there. You'll see a difference soon. As his body calms, he'll fall asleep more easily and sleep better.

Second, check your schedule. Has it been erratic? If your schedule is irregular, tighten it up. Get your child outside for the morning light and plenty of exercise, and turn off the television. By doing so, you'll set his body clock, making it easier to fall asleep.

Third, watch carefully and make sure that you're not missing the window of opportunity. You may assume that since he's napped, he's not ready for bed when he actually is. When you miss this "window," he gets a second wind that will carry him another forty to ninety minutes past his true bedtime.

Finally, review your expectations. When I asked Erica about Jack's schedule, she told me that he arose at eight o'clock a.m., took his nap from one o'clock to three o'clock p.m., and then was up until ten o'clock p.m. The total sleep time was twelve hours. His clock was perfectly set.

The trouble was that by evening, Erica was exhausted and ready for a break. So, we discussed potential solutions. "Would you consider changing wake time to seven o'clock a.m., so he'd be ready for sleep earlier in the evening?" Erica groaned louder. "That means I would have to get up, too." "Could you go to bed earlier, so it would be easier to get up?" "Hadn't thought of that," she responded. "Or would you allow yourself to take a break when Jack naps?" She smiled. "I like that one better. If I did, I would have energy to do other things at night. That would be nice."

Keep in mind your family's values and your child's needs as you plan for siesta. That way, it will be easier for you to honor it. When you do, the day takes on new possibilities. The evening doesn't deteriorate into an agonizing battle of wills. And, eventually, your child will go to sleep more easily—and, often, earlier—than you might ever expect.

Work together Ultimately, you want your child to be able to read his own body cues and to recognize when he needs rest. That's why it's so important to talk with him about how much better he feels after his nap or siesta. You can say things like "When we work really hard, we need a nap, so we can stay happy." Or "A nap helps us to stay happy at dinner time." Or "When we nap, we don't get so upset. Everyone is happier." Or "Naps help us to be very patient." When you teach your child the importance of rest, he may surprise you. That's what four-year-old Jared did.

"We were in the garden after lunch," his mom said. "I'd been working the past week to make sure he gets morning exercise. It changed our routine, because usually that's when I run my errands and he's strapped in. But I noticed that at school he ran around all morning and napped very well. So, I did it. I made sure he got exercise. I also started talking to him about how much better everyone felt when we had naptime.

"It took a week for it to make a difference, but yesterday, after lunch, he sat down on the ground and said, 'Mom, this is really hard work. I think I need a nap.' I dropped everything, mud and all, and put him to bed. He took a good hour-and-a-half nap. This is the child who used to scream if I even mentioned the word 'nap.' "

If your preschooler hasn't been taking naps, he's not going to be grateful to you for implementing them again. But I suspect you'll discover that even notorious nap-resisters can learn to appreciate the restorative value of naps.

Infants and toddlers are different Infants and toddlers need to stop, sleep, and consolidate their learning more frequently during the day than preschoolers. Often they hit their "window" much earlier and more frequently than you might expect, and, as they grow and develop, that "window" seems to keep moving. Keen observation is essential if you're going to keep up with them.

Finding the "window" for infants

Go back to the sleep journal you completed in chapter 8. Look for the signs of your child's "window." If your child is an infant, you may discover that only an hour to two and a half hours after awakening, he's ready for his nap again. If you can't see any signs, simply start noting his waking and feeding time, then attempt to prepare him for a nap two and a half hours later. If he screams, thrusts his arms, and is highly distressed, you know you missed his "window" and he's overtired, so, next time attempt to lay him down fifteen minutes earlier.

Continue reducing the time between awake and sleep times by fifteen minutes until you find the point where your child easily slips into sleep. This is especially helpful in the first months of your child's life, when it's unlikely that there is any clear schedule to his naps. Simply note when he wakes and eats, set your timer for an hour or more (depending on what you've learned about him), watch closely for cues, and get him down. Continue through the day as needed, carefully noting your child's pattern.

"We have followed your advice to watch carefully," Trish reported. "Amazingly, I have found his sleep schedule to be about every two and a quarter hours. I had heard that at his age it would be three and a half hours of wake between naps and bedtime, but that didn't fit him. On that schedule, he was extremely overtired all day. I never thought he needed it, his cues are really so subtle. It's just a yawn and a little eye-rub, but he still acts relatively alert, probably because he has adjusted to never being put to bed! Now I know when he's tired. But I notice that almost everyone else still misses it and doesn't believe me initially."

While conducting her study with Dutch families, Sara Harkness from the University of Connecticut told me: "Watching the infants, I noticed that when they began to fuss, I assumed they were bored and would have offered something to stimulate them. The Dutch parents, however, interpreted it as fatigue and put them to bed. The babies slept."

Dutch toddlers sleep an average of two hours more a day than children in the United States. Too often we choose stimulation when actually what everyone needs is sleep. What you will also discover as you allow your infant more opportunities to sleep is that the catnappers, who tend to nap only twenty or thirty minutes, begin to elongate their naps, because they are not overtired and on alert.

Watch for growth spurts

Felicity had been napping well and then stopped. It became extremely difficult to get her down. "What are we doing wrong?" her mom and dad lamented. Felicity had just turned six months old—a major developmental milestone. So, I asked them if they'd seen any changes in her behavior. "Yes!" they exclaimed. "She just started sitting up and rolling all over." Growth spurts, I explained, can throw off nap schedules, just as they do nighttime sleep. It's almost as though their brain is telling the babies that they are too busy to sleep. Checking back a week later, the parents were both grinning.

"We slowed down, relaxed, started watching her cues more closely, and helped her a bit more to settle. This week, she has been sleeping sooooo much, two- to four-hour uninterrupted naps and ten hours at night, waking only briefly to feed. Just when we thought we had her schedule all figured out, she throws us for a loop. We have now marked the nine- and twelve-month growth spurts on the calendar. We'll be ready!" An asset of a regular schedule is that when your child goes "off" and can't sleep, you know that something is up.

Finding the window for toddlers

The challenge with toddlers is that somewhere around twelve to seventeen months they begin to drop from two naps a day to one. Innocently, we may attempt to "push" them to one nap before they're ready. That's what Jessica, mother of one-year-old Becca, discovered.

Naptime for Becca was a nightmare. She'd get up at six forty-five a.m., and ride along to deliver her older brothers to school. After that, Becca and Mom ran errands. Near noon or one o'clock—the time varied—Jessica put Becca down for her nap. It didn't work. Becca screamed for at least an hour, and finally, in complete exhaustion, slumped into sleep for a twenty-minute nap. Then she was up and going again, but crabby, oh so crabby.

After learning about the importance of naps, Jessica made two decisions: first, to add more structure to Becca's day and, second, to watch more closely for her "window" for sleep. The changes were dramatic. Becca continued to awaken at six-forty-five a.m. and to go along to drop off her brothers at school, but instead of running errands, Becca was taken home. "I realized that Becca had been falling asleep for a short period of time in the car, usually about ten o'clock a.m. So, I decided to see if I could catch it."

At home, Jessica fed Becca a leisurely breakfast, then played with her and cleaned up a bit until ten o'clock. At ten, she told Becca, "Pretty soon, night night," and then nursed her. While Becca was still awake, she carried her up the stairs. "We'd go up real slowly. I'd hold her on my shoulder, rub her back, and say 'night, night.' Going slowly really seemed to help her. I'd lay her down in her crib, still wide awake. I couldn't believe it, the child who used to scream didn't."

Now Becca lies down and sleeps for two hours. It's a different little girl who awakens ready for lunch and a joyful afternoon. She catches a forty-five-minute to one-hour nap again around three thirty, and then goes right to bed at eight o'clock. Bedtimes are no longer an issue.

What happened? Becca wasn't ready to drop her morning nap, and by one o'clock p.m. was overtired. By watching closely, Jessica caught Becca's "window" for sleep. Their slow, low-key morning also kept Becca from moving into an overstimulated, hypervigilant state, so she could sleep.

"I realize now that I was trying to work her into my schedule instead of reading her cues," Jessica explained later. "Funny, she used to be so clingy, but when she gets her sleep, she's willing to go off and explore more on her own."

Even when your child stops sleeping, hold on to siesta Planning for and taking a break during the day is a great tension-buster. Learning to do it is a life skill that helps you stay in balance, which is why even if your child isn't sleeping anymore, maintaining siesta is still important.

By maintaining siesta, you also make the transition to no-nap more easily. While some children will suddenly announce one day that they no longer need a nap, most children gradually begin a process of napping one day and skipping the next. It's a very challenging time, with a child who is frequently overtired before the expected bedtime. But if you maintain siesta, your child has the opportunity to sleep if she needs it. That's what Molly found.

"I did the sleep chart and found that my daughter was sleeping twelve hours at night, so, I feel more comfortable with the fact that she's not napping. I have still kept siesta time right after lunch. On some days, she does nap, others she just plays, but now I don't get upset about it. If she doesn't nap, then I know bedtime has to be an hour earlier, so that she doesn't get overtired."

Siesta for teens Everyone benefits from siesta—even teens, who can learn to grab a few minutes between school and activities to center themselves and move back into calm energy. Think about it, it's likely that he's presently heading off to the vending machine for chocolate or caffeine-laden beverages, to recharge—why not simply stop and rest?

If, during the siesta, your teen does fall asleep, teach him to set a timer and limit the nap to twenty to thirty minutes. Doing so will make him more likely to awaken without grogginess, and the nap will not disrupt his bedtime. If, however, he is impossible to awaken after thirty minutes, or he feels very groggy, his body is letting him know he is deeply missing sleep. Together, make it a priority to review the demands of his schedule and help him to make different choices so that he can get more rest. It may very well improve his behavior, overall performance, and even your relationship.

Siesta for the family And the best part of siesta is that you can have it on the weekend for the *whole* family. Initially, in class, Amy reacted to this idea by declaring that her husband would never buy into a family siesta. But the next week, her eyes dancing, she declared it a success. "What did you do?" I asked her. She smiled. "I put the kids down and then I said to my husband, 'Honey, the kids are asleep, would you like to take a nap with me?' " I stopped her there.

Establish a naptime routine

The routine for naptime may be very similar to your bedtime routine, or it may be simpler. The key, just as it is at bedtime, is establishing and consistently implementing it so that the kids know it's time to sleep.

Transition activity When you're tired or frustrated, it's tempting to plop your child down for a nap without any transition, and then give up when it doesn't work. Just as they do at bedtime, children need to glide into sleep at naptime.

It's most helpful if your transition activity is a clear signal that it's time to stop, and begins to settle your child, like sitting down to eat, dimming a light, or reading a book. What's most important is that it's simple, and your child knows what's coming next.

I had decided to stop at Paidea, my favorite childcare center. Spring was being stubborn this year, teasing us with glimpses of temperatures in the 60s, then dashing our hopes with blasts of cold rain that slid into sleet. On this day, it threatened another bashing, and was dark even in the middle of the day, which was why I was surprised to find the lights off in the toddler room when I entered. I raised a questioning eyebrow to Angie and Amber, the teachers, who explained, "The intensity was a little high." I nodded and pulled up a chair to the table where eleven toddlers sat contentedly, eating their lunch. Angie and Amber sat with them, talking softly. "After lunch it's potty time, then naptime," they announced to the little ones. All nodded in acceptance, familiar with the clear transition routine.

Timing is just as critical as consistency. The closer you can be to the same time every day, the easier it will be for your child's brain to switch into sleep. "If we go to nap fifteen minutes later, it makes a difference," Angie told me. "It takes much longer for the kids to fall asleep. They're out of their routine, and it shows."

If your child attends a childcare center, find out what the schedule is there, and follow it as closely as you can at home. If possible, use the same transition activity. Your child knows it, and it will trigger his brain: time to sleep.

If you have more than one child, you can decide whether to attempt to move all of them to the same siesta schedule, or to stagger it so that you have time with each one. Once you've decided, keep it consistent. In class, Paula, mother of two, explained: "Initially, I was upset that each child was sleeping at a different time. I couldn't get anything done. But now I have a different attitude. I get time with each one. Adan would never let me read to Charlie. Now I can while Adan sleeps."

Connect and calm As Amber and Angie ushered the children back into their playroom after lunch, it was transformed. There was no doubt that this was a place to rest. The blinds had been drawn, dimming the room to near darkness despite the afternoon light. Toys were tucked away and cots set up. Blankets were laid out along with favorite teddy bears and pillows. A soft lullaby played on the CD. The rules were clear, too: now is the time to lie down. It's all right to choose a book, but only one book; after finishing it, there are no more. Then it's time to pull up the blanket, find the right position, and close your eyes.

Many children need help shutting out the world. Preparing the environment for your child in a way that clearly communicates to him that it's all right to stop and that nothing interesting is going on can be very calming.

Plan to let your child play quietly, look at books, or, if he needs it, offer a little back rub or, better yet, snuggle right up and take a nap, too. If he has difficulty winding down, help him. Watch carefully, and your child will show you what he needs. One child may respond to a neck rub, while another needs pressure on his back. Another may need to be nursed, or want a pacifier. A toe rub, back scratch, or a few pats may do the trick. Feel free to do what your child needs, just keep it simple and clearly communicate it's time to sleep.

Troubleshooting: Working with the nap-resister On the day I visited Angie and Amber's room, all of the children, except Matt, fell asleep easily. The spark was in Matt's eye while energy surged through his body. He didn't want to lie down. "Which animal would you like today?" Angie asked him. Matt ignored her. "Oh, a little off today," she said. Matt kicked his legs in the air. Angie stroked his chest. Matt flipped over and came up with his knees under him, his face in his pillow. Angie stroked his back while he was kneeling. She didn't force him into a prone position. Matt sat back up again. Angie picked him up along with his blanket and sat in the rocking chair with him. She gave him his pacifier, which had fallen from his mouth. She stroked his back as he sat in her lap, and sang him a lullaby. Finally, he began to settle. She laid him down then, lying right next to him, her arm stretched across his body to help him shut down, and continued to softly stroke his brow. When his eyes began to roll, she stopped, sat up, but stayed near. Within minutes, Matt was asleep. The message had been clear. It is naptime. I'll help you slow down, but it's time to rest.

Frequently, as your child is calming down to nap, about ten minutes into the process, there is a surge of energy. "I want to get up," he'll tell you. "I don't want to sleep." It makes you wonder whether or not he really needs this nap, yet intuitively you know that he does. Stay calm, remain firm, gently nudge him to sleep by giving him permission to let go and sleep. Let him know that he's safe and you are near, that it's all right to rest.

Establish a cue for sleep The cue for sleep at naptime can be very simple. It may be the same lullaby, music, or fan you use at night, or it may be a simple ditty. In class, when I asked the parents what their children liked, I learned that Ben looks at a book until his mother comes back in and tells him to stop and turn over because he sleeps better if he turns. A final reassurance that she is near and that she will check on him, is enough for him to slip into sleep.

Switch to sleep Just as you do at night, this is the time to allow your child to "switch" into sleep on her own. If she's not quite ready to do so, it's fine to work with her. Before two-year-old Michael was "switching to sleep" independently, his mother told me, "I would tell him, 'Your cousin Jesse lies down and goes to sleep on his own, and so does your friend Abby. Some day soon you'll be ready to try it,' and one day he was."

Ryan's mom found that she needed to rock him until he was drowsy. "If I put him in his crib, he'll stand right up again and get goofy or scream. Once he's asleep, which is usually in fifteen to twenty minutes, I can lay him down in the crib immediately. He opens his eyes for an instant, but then rolls to his side and closes his eyes again. I just have to make sure that he's on his way."

If stress is high, or you are *reintroducing* naps, you may find it's not the time to practice "switching to sleep" independently. In that case, you can say, "I'll stay while you sleep," and enjoy the break. After a week or two, when all is going well, you can begin to practice independent "switching."

Temperament: customizing the routine to fit your child

The teachers at Paidea put 125 children down for their naps every day. It takes them about twenty minutes once the children have moved to their cots. So, how do they do it? I wondered, too. "What do you think about when you're planning for naps?" I asked Carol, one of the preschool teachers. She thought for a moment and then said, "You have to figure out what each child needs to help him or her fall asleep, and then you do it. For example, Micah likes his cot in the corner, surrounded by walls on two sides. He enjoys the security of it and needs to have more space between himself and the other children, or he's tempted to talk. Joanna's cot is behind a divider. She can't sleep if there is anything to look at. Emma and Tyler both like their backs rubbed. I set their cots up side by side, so I can sit between them and rub both of them simultaneously. If I take ten minutes to do what they need, they get two hours of rest. I think that's a worthy payoff."

When your child is intense Hitting your child's "window" for sleep is critical for your intense child. Once this child is overtired, he can't self-soothe and is unlikely to be able to nap alone. And because he is naturally prone to being more vigilant, it is challenging for this child to stop and rest. You can expect that this is the child who will need a slow transition to naptime and extra soothing and calming activities. He may also benefit from having his nap "nest" be a place that feels like a safe cave or cocoon. It's worth the effort, because in order to manage his intense emotions, this child NEEDS his nap.

When your child is sensitive At night, this child requires that the environment be shut down, and that's also usually true for naptime. Removing interesting toys, dimming lights, adding a sound machine can all be helpful strategies to help this child shut out the world. That midday rest will also help him cope more effectively with high stimulation levels and a busy schedule.

When your child is slow to adapt Jim Cameron from the Preventive Ounce states, "Energetic, slow-adapting children are paradoxical: the more sleep they get at night, the more rest they can get during the day. The reason: Sleep at night restores their ability to handle frustration the next day. So, when they are asked to rest during the day, they are less inclined to protest."

While all children benefit from routine, it's especially important to the child who is slow to adapt. Routines let these children know what to expect, thus reducing their stress as they move from one activity to another. That way, they don't have to change both their expectations and their behavior.

A consistent nap site is important for this child. Unlike more adaptable children, who can nap wherever they may be, this child needs to be in his "spot" to be able to sleep well.

Remember, too, that if you are quick to adapt, it's more challenging for you to resist the possibility of another activity and stick to your nap plan. It may help for you to keep a journal in which you document your child's behavior on the days you stay on schedule and those you don't. I suspect the benefits of a regular routine will soon become so apparent that it will be worth the effort it takes to stick to your routine.

When your child's body rhythms are irregular Just as your irregular child may not appreciate your regular bedtime routine, it's likely that he will strongly resist a regular naptime too. Once again, however, he will benefit greatly from it. The more predictable it becomes, the more accepting of it he will be. One of the challenges, of course, is that an irregular child often has an irregular parent. If this is you, know that it's much harder for you to establish predictable sleeping and rest times for your child. Create a picture planner, ask a friend to back you up, or use your computer or watch to beep you and help you stay on course.

When your child's body rhythms are regular and predictable

I frequently run into the regular child who not only likes his nap right after lunch but also has a bowel movement then—which often wakes him before he's really ready to be woken. One-year-old Robert was one of those kids. You could have set your watch by Robert's body. He would fall asleep at noon and have a bowel movement every day at one o'clock p.m., right in the middle of his nap. Frustrated, his mother called for help. I wasn't aware of any way to change the bowel movement, so we simply moved his nap. Now he falls asleep at eleven o'clock a.m. and has a good two hours before the bowels start to growl.

When your child is energetic

High-energy kids hate to stop and take a break. The more excited and wound up they get, the harder it is for them to stop and rest. But a nap will restore this child's self-control AND prevent the accidents he's more likely to have. Helping him to "stop" is the key. Trish found a way to do it.

"Jason is so high energy that we used to let him walk around while we read to him at naptime. But now we go to the bathroom and then immediately get into bed. It makes a huge difference. I never attempted to rein him in before. Now I insist that he get into bed and slide under the covers. I tuck him in on all sides. I think it helps him to feel more secure and slow his body. I used to let him lie on top of the covers. When he gets excited, I remind him it is time to rest and that he needs to be under the covers. I stop reading until he gets back under. I tell him, it's all right to relax and be calm. Though he never says the words himself, it seems to allow him to do it. He also really likes songs. I sing and rub his back or his head. He likes that. When he starts to sing along, which only gets him excited again, I switch to a song he doesn't know."

Since naps were reintroduced, the change in Jason's behavior has been significant. Initially, when he missed his nap, he was all right for a day or two, but on the third day, he would wake up roaring, and continue through the entire day. He was aggressive with the baby, yelled at his parents, and got into trouble for squeezing and grabbing his sister. On those days, he was so wired that it was impossible to get him down for a nap. But now that he regularly naps, he's centered and calm. He's not perfect—but so much more open to guidance. Bedtime is also a breeze when he's had his nap.

Celebrating success Naps make everyone's day better. Naps really are good for relationships. And Bill discovered that with a little creativity, you can still have it all. "We wanted to have a birthday party for our one-year-old. All the relatives wanted to be part of it, but he really needs his naps. We came up with the idea of a 'brunch.' The party was after his morning nap and before his afternoon one. It worked beautifully. He was rested, thoroughly enjoyed himself, and the rest of the day went well, too."

It's easy to skip a nap, but, by evening, we are then left with nothing to offer our family. Their intensity and stress overwhelm us. There is no support or empathy to offer. The fatigue robs us of the joy they can bring to our lives. They lose out, and so do we, for a day without adequate sleep, with no time for reflection or rest, drains us and steals from us the rays of love.

PREVENTING POTENTIAL PROBLEMS

FIFTEEN

INFANTS

Getting off to a good start

"I wrote two thank-you notes for baby gifts, sealed them, stamped them, addressed them, and. . . . tried to put them into the disk drive of the computer."

—Ginger, mother of one

I sat on a rock overlooking a bay on the Kwishiwa River, in the Boundary Waters National Wilderness Area. I had paddled a canoe eight miles to get there. No motorized engines of any kind are allowed in this area, no boat motors, chainsaws, radios, or cars. On this day, as the sun rose, an eaglet perched in his nest atop a one-hundred-year-old white pine screeched for his breakfast. A loon called from the river to his mate, "Where are you? Where are you?" Soon she answered him, a lively trill, "I'm here. I'm here." I realized, sitting there, that the human infant was actually designed by Mother Nature for a world like the boundary waters, but in today's society it's more likely that we will find him being wheeled through the aisles of a Super Wal-Mart. It's a BIG shock to his system.

Marie Hayes, a researcher from the University of Maine, states, "One of the things that is under-recognized in the child development literature is the continuity from prenatal to postnatal stages of life."

After birth, your newborn is challenged. Everything is different. While older children may be attracted to "new" things, the newborn seeks familiar cues. It's during those first few months after birth that he is figuring out what are safe areas. Mom is the infant's beacon in making the transition from prenatal to postnatal. The adopted child will home in very quickly on the nurturing actions and sensations associated with his adoptive parents. What the newborn intuitively knows is that his caregivers are the source of food, heat, touch, and, most important, familiarity. Cues associated with his caregivers, especially

Mom, are easily recognized. He is conditioned to tastes and odors linked to her. He even recognizes her voice and prefers the intonations of her language. Being close to her means safety to him.

That's why, in order to get your baby off to a great start, it's critical to remember that what he likes is that which is familiar to him—the sounds he heard in utero, the smell of his mother, confined space and movement. What he does not seek is "new," overwhelming sensations or stimuli.

But rather than feel comfortable offering the closeness and care for our newborn's needs, we often worry that we may start "bad habits" or "spoil him." What I'd like to offer you instead are two goals for those first few months: first, to sensitively ease his transition from the prenatal to the postnatal, and, second, for you to discover who has come to live with you.

What's normal?

If you stop and think about what your baby has experienced in the womb, it's easier to recognize what you might expect from him during the first three months at home. The womb was dark, so he doesn't yet know about night and day. In the womb, he slept. But it didn't matter whether his mother was moving or lying still. No one discussed "consolidating" sleep into manageable stretches with him, and there was no need to do so. He never felt hungry; it's a new sensation to him, and often quite unnerving. He never felt cold. That, too, is quite shocking and upsetting. Sounds were muted. He could hear them, but there was always a buffer. Colors and bright lights are all so novel that he may have difficulty disengaging from them and can easily become overwhelmed. Even the sensation of his arms flailing or legs kicking freely is disconcerting. And if that's not enough, he's working on doubling his weight in the first few months. He has much to learn and a great deal of growing to do. It's all exhausting.

If you think about it this way, it's not too surprising then to learn that it's typical for a newborn to average fourteen and a half to eighteen hours of sleep a day. There are significant individual differences, however, and the "range" is actually nine to twenty-one hours. In the first few days after birth, the baby still has hormones from Mom that can help him sleep. After a few days, those are gone, and the baby is not yet producing his own. As a result, there is rarely a predictable pattern to his sleep. Instead, he sleeps, wakes, eats, checks out the world around him, seems to say "oh my," and falls asleep again, eager to store all of that new information in his brain.

"After Jackson's birth, he'd sleep two or three hours at a time and then would be up for thirty minutes," his mom told me. "I'd feed him and change him, and he'd go right back to sleep. He was probably sleeping nineteen hours a day."

Each baby seems to have his own pattern of sleeping and awakening that is not tied to his size or gender. Nor is his internal body clock responsive to light or dark until he is four to six months old. Initially, his cycle is more related to when he's hungry.

Researchers have demonstrated that there is a predictable progression of maturation in sleep with increasing periods of wakefulness and longer periods of sleep during the night. Sometime around six to eight weeks of age, you can expect that the baby, who was waking every two or three hours, will begin to sleep for a block of four or five hours. Dr. Tom Anders from the University of California, Davis, found that 44 percent of two-month-olds and 70 percent of three-month-olds were sleeping for a stretch of five hours. However, those same infants often were awakening more frequently during the second half of the first year as new developmental skills and stress in their environments alerted them. Only 6 percent of two-month-olds and 16 percent of nine-month-olds were sleeping from the time they were put to bed until they woke for the day.

When you look at how much work your baby has to do in those first few months of life, it's quickly apparent how important it is to establish a "nest" for him where he feels safe enough to sleep.

WHERE SHOULD A BABY SLEEP?

A friend once asked me how I had decided to go to college. I remember looking at her with a blank stare, realizing at that moment that it had never been a "decision." While neither my mother nor my father had attended college, they both highly valued education. In our family, it was simply assumed that my sisters and I would graduate from high school and go to college, and that's what all five of us did.

In most cultures of the world, where a young infant should sleep is not a decision. As with my experience of going to college, it is simply an assumption handed down through generations. A young infant will sleep with his mother or in very close proximity to her, in a cradle. It's only in the last two hundred years that we've challenged that assumption. Where an infant will sleep can be a decision that may be excruciatingly difficult to make, but it doesn't have to be that way. You really can choose what fits your family best and what is safest for your infant.

Creating a nest for sleep

Humans are mammals, and mammals like their "nests." The more stable and predictable your infant's "nest" the better he will sleep. The key with the "nest" is consistency for the first three months. At the end of this period of time, there is a significant shift in biological rhythms, and if you choose to move your baby to another site at that point, it can be relatively easy to do.

Fortunately, there are many options to fit our individual needs and lifestyles. The most important factors are: the site needs to be a place where you are comfortable having the baby sleep AND that when the baby is there sleeping, you will be able to sleep, too. If you sleep best with your child tucked next to you in bed—you'll probably choose to co-sleep. But if having a tiny person next to you keeps you wide awake listening to every breath, you'll probably want to choose an option that places her either in a bassinet or a crib in your room (which is what the American Academy of Pediatrics recommends), or in a crib in her own room with a monitor.

I'll let you decide what option works best for your entire family. If you are parenting with a partner, he/she will also need to be in agreement. Remember, a baby senses her parents' stress. If you are battling over where the baby is sleeping, no one is going to sleep well. Together, discuss your needs. Think about your temperament. Is one of you a light sleeper while the other could sleep with three kids in bed with you and never notice? That matters. What's your own experience? Did you sleep with your parents, or was that taboo? Then be ready to change your decision after your baby is born, because it's likely that your infant will want a say in this, too. There's also nothing like holding that little miracle in your arms to bring out a protective furor you never imagined.

As you think about where you want the "nest" to be, make sure it is a location that you are comfortable using for naps as well. It's also important that you can feed your baby either in it, or near it, since we can predict that he's going to be up and wanting to eat. You'll want to avoid feeding him in locations where you don't want him to sleep. Remember, it's those olfactory cues and tastes that let him know, "This is my nest. I'm safe here and can sleep." If you do not want to co-sleep, for example, either avoid nursing in your own bed completely or, as soon as you finish, move your baby into his sleeping space. The more he sleeps in his "feeding space," the more he'll enjoy sleeping there.

Choosing to co-sleep Co-sleeping means that your baby is sharing a bed with you. The American Academy of Pediatrics does not recommend babies sleep with their parents, but James McKenna, director of the mother-baby sleep laboratory at the University of Notre Dame in Indiana, and breast-feeding groups disagree with this recommendation. The controversy continues. If you do choose co-sleeping, remember, Mother Nature designed this style for simple bedding, frequently on the floor, without crevices or creases to be trapped in. No feather beds, thick pillows, bumper pads, or high mattresses that babies can roll off of, no water beds or heavy blankets. So, if you choose this style, consider taking your mattress off the frame and putting it on the floor, or at least on a frame that is close to the floor, get rid of excess pillows, and keep the covers minimal. Maintain a cool room temperature, about 68 degrees, so that everyone can sleep comfortably, and make sure the mattress is big enough so that you can move without worrying about rolling over the baby. Do not sleep in recliners, couches, or chairs.

The nice thing about co-sleeping is that it's convenient to simply roll over and nurse the baby in the middle of the night. The smell of Mom is everywhere, because she's there, and that means safety to the newborn. He will still wake, but it's simple to comfort him without crawling out of bed. The convenience minimizes your arousal and makes it easier for you to return to sleep.

Co-sleeping is also very portable. Whether you're at home, visiting Grandpa and Grandma, or staying overnight in a hotel, baby can rest surrounded with the cues that tell him he's in the right place, because he's near Mom.

The drawback to co-sleeping is often the worry that if you let your child in your bed you'll never be able to get him out. Unlike the Indian culture, which allows children to choose when they move out of the parents' bed, which is often at puberty; or the Japanese approach, which promotes the shift when the child starts school, few American parents wish to have their children in bed with them for an extended period of time. That's when it's important to remember this choice is initially only for three or four months. Then you can decide to change it, or continue. It will be up to you.

Avid co-sleepers love their arrangement, and parents find no problem locating private places in their home for intimacy. As one dad said, "Having kids in bed with you is no excuse for not having a satisfying sex life. If the kids are all in your bed, it means that every other room in the house is available. It adds a little spice to life."

Co-sleeping may be a perfect choice if you are an employed parent who is separated all day from your infant. And researcher James Mckenna from Notre Dame has found in his work many benefits for the baby. In his studies, he found that co-sleeping babies match their mom's breathing, and the proximity also seems to help them to regulate their body temperature and heart rates.

When not to choose co-sleeping The concern about co-sleeping is the potential of "lying over" or smothering your baby. Be aware that there are certain factors that reduce your sensitivity to where your baby is in the bed. So, do NOT choose to co-sleep if:

1. You smoke

2. You consume alcoholic beverages or use drugs, including prescription or recreational or over-the-counter medications, which may impair your responsiveness

3. You are retaining significant weight, which may hamper your awareness of where your body is in relation to your baby

4. Your partner doesn't want to—babies need parents who are calm and working together

Sleeping with your baby near but not in your bed If you want your baby near but worry about rolling over on him, find yourself always on "alert" if he's next to you, or simply are not comfortable co-sleeping, the sleeping "near" may be your best option. There is a wide assortment of port-a-cribs, cradles, hammocks, and bassinets that make it easy to keep your infant near you but not in bed with you. You'll want to choose something that is easy for you to use both at night and during the day for naps. It's also helpful if it's portable, so, when you travel during those first few months, you can take it with you. But even if it's not, we have other methods for "taking the nest along."

"Our son slept in a cradle right next to our bed," Jeanne explained. "In the middle of the night, I just grabbed him and nursed him. I didn't even get up. Then I'd burp him, lay him down, and say good night. He went right back to sleep."

Expect to transition your child into a crib or larger space as he grows. That's what Kelly did. "As newborns, my kids all spent the first five months in a bassinet in my room. Then they moved to a crib. I started with naps in the crib, and then moved them. They slept in the crib until they were two, and started climbing out. When they were toddlers, I had them nap on a big bed, so when they climbed out of their crib, they had already been sleeping in a big bed and went right to it."

When you create a "nest" separate from your bed, it's important to bring in the smell of the nurturing parent. An easy way to do that is to purchase a supply of burp cloths or small, light blankets, all the same color, texture, and size. When you feed your baby, throw the cloth over your shoulder and chest, so it picks up the smell of your body. When your baby is almost asleep, or has just fallen asleep, lay that cloth in his "nest" first, and then lay him on it. (Be sure the blanket is not heavy or bulky, to avoid smothering.)

Every time he naps or goes down for the night, go through the same routine, laying him down on his burp cloth or blanket infused with the smell of Mom or the person he is most familiar with. If you're traveling, take along your "burp cloth": it makes your child feel as though he's brought his "nest" along. If he's a highly sensitive sleeper, take along his sheet and mattress pad as well.

When we talked about the "burp cloth" or blanket, Leah started to laugh. Shaking her head, she admitted: "I still have remnants of my blanket. I took it to college with me even when I studied abroad. I remember once someone asking me why I still had my blanket at twenty-one years of age. But I could take one whiff of it and know everything was all right. I had my 'nest' no matter where I was."

Jeanne looked amazed, and then confessed, "Mine went with me to labor and delivery for both of my kids!"

Baby sleeping in his own room, next door, or down the hall If you are a light sleeper and cannot sleep with your baby in your room, it's important to recognize it. You need your sleep, too. If this is the case, then you may choose the option of your baby sleeping in a separate room. (This may feel most comfortable to you for other reasons, too.) If this is your choice, prepare the room with a crib, baby monitor, comfortable rocking chair for feeding, and, if there's room, a bed that you can sleep in when needed.

The burp cloth or blanket comes to the crib as well. When you feed your baby during the night, use the rocker in that room, or bring him into the adult bed in that room and feed him there. Then return him to his crib. Inconsistency in sleep site causes infants to become aroused. Predictability and stability allow them to remain calm, comforted by the smell of Mom, and ready for sleep.

No matter what site you choose for your infant's "nest," remember to select a site that's easy for you to use and allows you to remain as calm and rested as possible. That way you'll be able to provide your baby with the sensitive care he needs to know that he will be responded to, and that it is safe for him to sleep.

How to get your baby to sleep In Holland, only 9 percent of babies are considered "fussy." In the United States, that number is 19 percent. The difference, many believe, is in the routine. Dutch families begin a routine from day one. It lowers stress hormones for the mom and, as a result, for the babies as well. Creating a routine does NOT mean putting your baby on a rigid schedule. What it does mean is creating soothing, calming strategies that allow everyone to be ready for sleep. The way to begin it is by working together.

MANAGING THE TENSION

Easing parental stress—finding ways to reestablish intimacy

The moms in my class called it "touched out." After nursing and nurturing a young infant, the last thing they wanted was to be "touched" by their partners. Instead of feeling pleasurable, the touch was experienced as invasive and irritating. It left them feeling guilty and put a real damper on their sex life. But when I asked what their partners needed, Natalie quipped, "Sex." "Let me put that a different way," I replied. "What does your partner need to feel loved?" "Sex," Natalie repeated, giggling. I tried once more, "What does your partner need to feel special?" "SEX!" the group yelled in unison.

Mother Nature strives to space babies for the survival of the species. That means she increases the hormones that lower the sex drive in new moms. Unfortunately, that reduction affects only one side of the equation; the dad maintains a strong sex drive, which can leave him feeling as though he's been left out in the cold.

Reestablishing the couple relationship after the arrival of an infant is a crucial task in the first year. Jay Belsky, a professor at the Institute for the Study of Children, Families and Social Issues in London, found in his studies that the more involved dad is with the infant's care, the more positive feedback he gets from the baby and the less stressed the mom is. A less-stressed mom has more time and energy for the dad, which makes him happy and willing to help out more. By working together, you can not only ease the stress, but enhance your relationship as well.

That can mean for moms letting go of "your way" of caring for the baby and allowing dad to figure out "his way." Remind the other women (grandmas, sisters, friends) in your life not to tease or take over from dad. Often, unintentionally, women can bond together and exclude new fathers.

If you're breast-feeding, make "bath time" dad's job. It offers a wonderful, intimate opportunity to care for the infant. Dressing, changing diapers, carrying, rocking, and putting baby to sleep after he's eaten are all tasks that dads can take over. Studies demonstrate that children whose fathers are actively involved in their care the first six months of life score higher on subsequent measures of intellectual and motor development.

And when it comes to reestablishing intimacy for the couple, it's important to remember that intimacy is a genuine feeling of love that can be communicated in many ways. Obviously, sexual intercourse is a profound message of love, communicated through tender and thrilling touch. In an enduring relationship, however, it cannot be the only way of communicating one's deepest and most caring emotions. Intimacy is a message that must be communicated intentionally in multiple ways, so that it is not lost in the demands and rush of daily life. While it may seem hard to believe at times, it really is possible to be intimate and be a parent.

As one mom tried to explain, "The first task is to get the engines running again." They're a little hard to start after a new baby, a bit like starting a car that has set out on a below-zero night. But the battery isn't dead; it's just a little sluggish. What's important to help your partner recognize is that while "Hey honey, how about tonight?" may be an enticement, even better may be "Why don't I do the dishes so you can take a bubble bath?" Or "Would you like a back rub?" Or "Why don't I bring home dinner tonight?" Or "I'll get up with him tonight so you can get a little extra sleep."

The invitations need to go both ways. But as you care and nurture your little one together, don't forget one another. Establishing a routine of shared care and intimacy will make you happier and less stressed. Your child will feel the difference and everyone will sleep better.

Easing baby's stress

When new parents are feeling good, baby's stress level is lower. Once you've gotten that in order, there are a few more things you can do to help him pass through the transition from womb to world.

Swaddle your baby Your newborn is used to cramped quarters. The freedom of waving his arms and legs can be overwhelming and overstimulating. That's why it's so helpful to keep him swaddled. Babies need to be laid on their backs to sleep. It's a preventive measure that drastically helps to reduce the risk of Sudden Infant Death Syndrome, but it also leaves babies more likely to startle if they're not swaddled.

In order to swaddle your infant, you'll need a blanket large enough to encompass his body. Begin by laying the blanket out as a diamond. Place his head about twelve inches from one corner. Fold the corner at his feet up to his belly, and then fold the corner on the right across his chest, placing his arms in the cross-chest position of the womb. Take the left corner and firmly wrap him. Tuck the end into the fold. Now he's bundled like a papoose. This position, rather than feeling like a straightjacket to him as it might to you, is actually familiar and comforting. Continue swaddling him until he kicks himself out of it and lets you know he's ready for more freedom.

Often, if your baby is too sleepy to nurse, it's recommended that you unwrap him while feeding, in order to arouse him enough to eat. Watch your baby carefully to ensure that this arouses him without overstimulating him. Does he open his eyes wide and look at you calmly, slowly cycling his arms and legs, or does he become stiff and agitated, looking away from you because he's overstimulated? Avoid unwrapping him if the freedom overwhelms him.

Use nightgowns The pajamas are cute, but remember, your baby finds it stressful to be cold. Try dressing your infant—girls *and* boys—in a nightgown at bedtime. "A nightgown allows you to quickly change your baby during the night," Pam Kennedy, whose career was spent in hospital nurseries, told me. "You just pull it up, switch the diaper, and pull it down. It's easy, and allows you to avoid pulling on limbs. The exposure to cool air is minimized, and, as a result, arousal during diaper changes is less likely to occur."

The other nice thing about a nightgown is that it often comes with hand covers as well. Use them. They, too, can be comforting to your baby, who is not used to seeing his own hands waving in front of him, and cannot control his fingers enough to stop from scratching himself.

Try a cap Often, in the hospital nursery, your child is given a cap. This isn't just for looks. Medical personnel have discovered that infants sleep better when they're warm—but not overheated. Babies lose a huge amount of heat out of the top of their head. A cap not only keeps them warm, but seems to comfort them. So, try using that sleep cap for your little one. It will keep him warm and help him sleep.

Celebrate thumb-sucking If your baby sucks his thumb, there is reason to celebrate. Thumb-sucking babies sleep better, because they can soothe themselves. I know there's a concern that doing so will mean you will one day end up with an orthodontist bill, but in my experience, there is *always* a reason for orthodontia, whether or not your child sucks his thumb. You might as well enjoy the benefits while they exist.

Massage Rocking is very soothing for babies, but stopping it may lead to them awakening. Massage is even better. Dozens of studies document a decrease in crying and lower levels in stress hormones in infants who have been massaged. Take a few extra minutes after your baby's bath to massage him. While you are feeding or holding him, stroke his legs and belly. Whenever you can, TOUCH your baby. Carry him in a pack against your body. Avoid leaving him in hard plastic carriers, unless he's traveling with you in his car seat. He likes to be "molded" to your body, not to plastic. You will not spoil him, you will soothe him. A happy baby will one day reward you by being more patient and open to your guidance.

Shhh, shhh, shhh

It may sound silly, until you think again about the womb and the sounds your infant heard there. The whoosh of blood and the beating of the heart are familiar to him. By making a *shhh, shhh, shhh* sound and gently swaying your baby back and forth while he lies on your forearm, you remind him of his "old" home. It's a sensation that feels safe to him and helps him sleep.

Cueing for time to sleep Young infants are not ready for a schedule for sleep. During the first few months, it's important to go with your baby, noting when he needs sleep, when he's ready to eat, even if his timetable is uniquely his own. What you can begin to do, however, is to introduce to him the difference between daytime and nighttime.

The young infant's brain is not yet impacted by day and night cycles, so, he has no idea of when you'd like him to be sleeping. While you will protect him from noise and light that is overstimulating, do expose him to the normal noise of your day. When he awakens, care for him and talk with him. Let him know that during the day you do like to play with him.

It's important, however, to maintain a pace that's low-key and slow enough to let you observe the cues that he's ready for sleep and to allow him to get it when he needs it. Most American infants are averaging ninety minutes less sleep a day than standard recommendations. If you sense he's bored, try putting him to bed rather than stimulating him.

At night, keep lights off or dim. Minimize interaction, communicating to him that nighttime is a time to sleep. If you have older children, include him in the transition and wind down to bedtime. While he won't yet be ready to follow this schedule, he'll begin to learn the cues. Even if he's your firstborn, establish a bedtime routine. By the time he is six months old, a more regular pattern will begin to appear.

When you can, try laying him down while he's still drowsy and giving him the opportunity to "switch" to sleep on his own, but do NOT leave him to cry. If he's falling asleep nursing or in your arms, don't worry about it at this point. Recent studies indicate that there is NO correlation between self-soothing and better sleep habits in the first four months of life. And some researchers are even questioning the correlation during the first year. Ultimately, however, this will be true. In order to prepare him, you can try practicing with him, gradually introducing him to the idea of falling asleep on his own.

"When Tara was a baby," Mike told me, "her cues were easy to read. She'd rub her eyes and roll her head to the side. Every night we would dim the lights, change her, and put on her nightgown. My wife would nurse her to sleep and, as she did, I would turn on the same classical song. Every time she fell asleep, the song would be playing. By five months, she wasn't always falling asleep at the breast, so I would take her, turn on the music, and lay her down. Now she's fourteen months old, and when I take her upstairs and go through our routine, she lunges for the crib."

Taking time to note your baby's cues, repeating similar steps, and creating a little signal begins to introduce your baby to the cues that soon will help him know it's time for sleep.

In the meantime, enjoy cuddling him and recognize that since you cannot adjust your newborn's schedule yet, you can adjust yours. If you know that he will sleep three or four hours beginning at eight-thirty or nine o'clock p.m., consider moving up your bedtime so that you allow yourself a stretch of sleep before you're aroused again. While it's tempting to use that time to catch up on work or things around the house, your own sleep and well-being these first few months are critical to your ability to cope with the demands and stress of having a newborn.

Customizing for temperament

Three-month-old Kyle has found the perfect time to party with his mom—three o'clock a.m. The house is quiet and dark, and his two older brothers are finally silent and sound asleep. He lies wide-eyed, gazing at his mom, happy to be near her. If he can't see her, he screams. His "party time" can last up to two hours, and his mom is exhausted. "What can I do?" she groaned.

After watching Kyle carefully, I realized he was a highly sensitive baby, noticing every sound and movement in his environment. "I suspect he wakes ready to play, not to torment you but because the stimulation levels are finally low enough for him to be free to interact and take in his world," I said. "It's true," his mom responded. "If I stay home and let him get his naps, he has a better day and night."

When it comes to sleep, even the newborn demonstrates temperamental differences.

If your baby is intense The nurse in the hospital nursery may have been the first to notice your child's intensity. It's not unusual for the intense child to be sent to Mom with the explanation that his screams are waking all the other babies in the nursery. That's after he'd been sent to the nursery to give Mom a break. It's the intense baby who strives to be near you. He seems to know that your closeness will help him to keep calm. It's important to remember that it's all about physiology. He needs you to reduce the stimulation in his day, protect his daytime sleep, and, when he cries, to help him slow his heart rate by holding and rocking him.

This child also tends to like to suck. Sucking calms him. If you are nursing, you may worry that you do not have enough milk for him, but the reality is that he is often simply soothing himself.

It's also this child who will take a little longer to learn how to "switch" to sleep independently. At seven months, Noah seemed like a rather mellow child, who could be laid down and go to sleep, but Mom told a different story. "Initially," she told me, "I'd watch for his cues, like sucking his thumb or a whiny kind of cry. I'd rock and nurse him then, and lay him down. He'd scream. I would pick him back up. Otherwise, he'd be wide awake again. I'd walk him then, until he fell sound asleep. I repeated this process every day for three months, just to let him 'practice' a tiny bit. One day, he didn't cry when I laid him down. He just looked at me. I said, 'Night, night, and I'll be here.' He lay there, looking at his mobile, and then stuck his thumb in his mouth, flipped to his tummy, and went to sleep."

The intense baby needs a sensitive, quick response that prevents him from becoming overwhelmed, but that doesn't mean you never practice with him. Keep the practice session minute. Respond immediately if he is distressed by the attempt. Be patient, and allow time, and he, too, will get there.

In a classic study completed in 1984, researchers found that the intense infants were more likely to survive a famine in East Africa than their mellower counterparts. They insisted on getting adult attention. So, celebrate that you have been sent a "survivor" baby.

If your baby is sensitive Like the intense child, the sensitive baby needs "protection" from overstimulation. Nothing gets past him. This is the baby who hears the click of the door and the creak of your knees as you stand to lie him down. Pour a cup of coffee, and he looks up as the liquid hits the cup. He turns toward the fleeting shadow, or the movement outside the window. Sharp and perceptive, his little body often stays on "alert," and sleep eludes him.

Your highly sensitive child will let you know he's overstimulated, because he won't be able to sleep, or, like Kyle, he'll wake in the middle of the night prepared to play with you. He often does best if you "wear" him in a soft carrier. Avoid stores with overly bright lighting. Turn off the television as "background noise." Limit the mobiles and other toys around his "nest." Make an extra effort to manage your stress. Swaddle him and massage him. Gradually, he will find it easier to shut out the world, but for right now he needs a little assistance from you.

If your baby is high-energy Safety is a real concern for the high-energy infant. Even newborns that are temperamentally active can crawl across an adult's bed. If you use an infant seat, be sure to set it on the floor rather than a table or counter. Use the safety straps. Do not leave this child where he might scoot or roll off. If you've co-slept with this child the first three months, you may choose to move this little one to his own bed—unless, of course, you don't mind little toes in your face.

Celebrate successes

The card arrived in the mail with a little brown envelope attached to it. "United States Postal Service postage due," it read. I opened it and found a thank-you card from a friend who had recently delivered her fourth child. "Thank you so much for the outfit," she wrote. And then, because we'd been chatting about sleep, she added: "He still hasn't worked himself into any type of rhythm, so, sleep deprivation is an issue around here. When I'm up with him, I'm trying to focus on the beauty of the night, because that's the only quiet around here."

What an amazing mom, I thought. Even in the fog of her exhaustion, she has found a positive focus. And then, as I slipped the coins into the brown envelope for my carrier, I had to laugh. Forgetting the stamp was such a blatant reminder that sleep deprivation takes its toll even on the strongest.

Give yourself credit. Being the parent of a newborn is VERY hard work. Take a break from other outside commitments. EXPECT to be awakened during the night. ASK for help from those who love you, and choose to sleep whenever you can, day or night. Then you can relish the sounds of the night as you care for your baby. Watch the moonbeams as they fall across his cheek, breathe in deeply his "newborn" smell, and cherish the miracle that has come to live with you, knowing that these days shall pass much more quickly than you ever imagined.

SIXTEEN

TAKING THE FIGHT OUT OF THE MORNING START

Adjusting wake and bedtimes

"I'd like to leave a wake-up call for whenever it is I wake up."
—The Family Circus

You could set a clock by little Helen. No matter where she was or who she was with, at seven o'clock p.m. Helen went to sleep. She didn't even need a bed. A couch, the floor, just about any relatively comfortable spot was adequate. In many ways, this delighted her parents, who found that her ability to fall asleep anywhere, and the predictability of her body clock, gave them a great deal of flexibility. However, Helen's clock also was finely tuned for a five o'clock a.m. wake time. This did not make her parents happy, and they came to me, asking how to help Helen with her sleep problem. The challenge was that Helen, who also took a two-hour nap every afternoon from twelve-thirty to two-thirty, was getting just the right amount of sleep for her. Unfortunately, her schedule didn't quite fit with the rest of her world. So, what to do?

Adjusting schedules affects the circadian rhythm

There are many reasons for changing a child's schedule. It could be as simple as a slumber party or a short trip across a time zone that temporarily interrupts the system. Or it may be necessary on a longer-term basis to cope with a shift to Daylight Savings Time, a move across the country, a return to school after summer vacation, or to ensure that your child gets the sleep she really needs. And then again, it may be because you simply want to do something fun.

Whatever the reason, helping your child to be flexible enough to cope with everyday life is an essential skill. What's important to remember, however, is that while adjustments can be made, and often need to be, it isn't merely a matter of changing the time your child gets up or goes to sleep. Altering the schedule affects the circadian rhythm, the body clock that tells us when to be alert as well as when to release various hormones in the body, adjust body temperature, pass bowel movements, and much more. When your child is feeling well rested and awakening on his own, he's functioning in sync with his own circadian rhythm. All systems are aligned. He has a positive attitude, and his energy is high.

But when we change the schedule and have to awaken him before his body is ready, or hold him off until we want him to fall asleep, we are disrupting an entire physiological system. That's why moving sleep or waking times even an hour can require two to three weeks of adjustment until your child is once again in tune with his circadian rhythm. It is an adjustment that needs to be approached thoughtfully and with a plan in mind.

MOVING WAKE AND BEDTIMES LATER

Lydia and James were at their wits' end. Their four-year-old son Jesse would go to sleep each night at eight o'clock p.m. and sleep straight through until five o'clock a.m. Drumming his fingers on the table, James told the group, "By 5:05 a.m., he's up, raring to go and expecting that everyone else should be, too. He's very extroverted and energetic, and wants someone to get up to talk with him and play. We are not morning people. About seven o'clock a.m. is the earliest we can muster any energy, and with an eight-month-old baby who is still waking in the night, seven o'clock a.m. is pushing it.

"We've tried making him stay in his room. He throws a fit. We attempted to teach him to read a clock, but he comes running into our room, gleefully shouting, 'The clock says five o'clock!' And when we explain that it doesn't say seven o'clock a.m., which means that he is supposed to go back to sleep, he unplugs it and throws it against the wall. Even if we let him in bed with us, he plays with our hair, tickles us, or talks. We need some help."

Managing the tension

Changing your perspective I have to admit, I am a morning lark. So, when Lydia and James first asked their question, I couldn't help thinking about how much I enjoy mornings. Immediately, I jumped into a monologue.

"One of the most delightful things about having children is that they will expose us to experiences that we may never have without them!" I exclaimed. "It may be a view of the stars as we feed them in the middle of the night or, years later, wait for them to come home after a date. Then again, it may be an opportunity to see the beauty of a sunrise."

I paused for a short breath before asking the group, "What are the benefits of having an early riser come to live with you?"

Harry, another morning lark, laughed out loud as he remembered fondly the early years with his daughter. "You couldn't resist her," he chuckled. "She'd wake up grinning at five o'clock a.m. and nothing could convince her to go back to sleep. Finally, I acquiesced. I was awake anyway, so I started getting up and having breakfast with her. Fortunately, my hours at work were flexible, so, I decided I might as well start my day early. I left the house at six-fifteen, when my wife got up. I completely missed rush-hour traffic and had over an hour of quiet time to get things done before others started arriving. It also allowed me to leave earlier and enjoy family dinners. Even after she started sleeping later, I kept the schedule."

Lynn jumped into the conversation. "Brandon is my third child. In the morning, it's quiet and he's so cuddly. I just go to bed early myself so that I can treasure every minute of those early hours. I know they aren't going to last long."

My eyes twinkling, alive with the glory of early morning, I went on to explain to the group that early risers often have a genetic tendency to rise with the sun. When they do so, it's because their body clock tells them this is the best time for them to be alert. It's not a plot against their parents. We don't know why, but male children are more likely to be early risers. Springtime also tends to jump-start the morning larks as much as it does the birds. Falling back from Daylight Savings Time often creates early risers, as they lock into their old wake time. "If you have an early riser," I said, "you can choose to look at it from a new perspective. Maybe you're raising the next top morning disk jockey!" And then I stopped long enough to see the look on James's and Lydia's faces.

James snorted, almost a low growl in his throat. "I don't want a new perspective. I don't care about beating rush-hour traffic. I don't want to go to bed early. Sometimes I WANT to watch Jay Leno. I need him to sleep!"

Realizing I'd completely missed the point of the question, I grimaced and changed my tune. "Sometimes," I said, "the mere idea of waking and actually getting out of bed at five o'clock a.m. is absolutely abhorrent. It just doesn't work with your family. The good news is that when that is the case, you can tweak it." James nodded in appreciation.

Check the tension level throughout the day Combine a "morning gene," a temperamentally energetic child, add a little tension, and you have an early riser. While you cannot change his genes, you can manage the tension that tends to trigger those genes, by putting him on alert. Review your sleep journal. On the mornings that your child arises early, check what happened the previous day. Here's what the parents in my group discovered.

"Inga wakes up earlier when the previous day has been hectic and rushed."

"Sam's up when he hasn't had enough exercise the day before."

"If she's excited or anticipating an event, Tara will wake at the crack of dawn."

If you can predict it, you can plan for success by taking extra measures to soothe and calm, so that when your child goes to sleep, his body isn't wired. A relaxed body experiences deeper and longer sleep. But if that's not enough, it's time to move the "clock." I'll begin by telling you how the sleep scientists recommend we move it, and then I'll tell you how others move it when they forget to do or can't do what the scientists recommend.

Time—resetting your child's body clock

Begin with breakfast and lunch Your child's brain not only tells him when to be alert, but also when to eat and to eliminate. That's why, when you want to move his clock, you have to move the entire "system." Begin your tweaking process by delaying breakfast, lunch, and dinner fifteen minutes. That's it, fifteen minutes. The body clock moves slowly, so, we have to move it in increments over a period of days or even weeks. You are shifting an entire physiological system, and you are balancing between gently nudging your child and yet not wanting to push him into a state of overtired.

Don't forget naptime The common advice for an early riser is to skip the nap, so that she is more tired and will sleep longer. The problem with this advice is that it does not consider the early bird's strong body clock, which tends to awaken her no matter how tired she is. Lack of sleep also creates tension, which puts her on alert. In addition, if she skips her nap, she'll want to fall asleep earlier, which can actually lead to an even more untimely rising. Instead of skipping it, make sure your child gets her nap, just move it fifteen minutes back, as you have done with her meals. That way, she'll be rested enough in the evening to delay her bedtime fifteen minutes without getting a second wind.

While your child needs her nap, do adjust its length. After reviewing their sleep journal, Stephanie discovered that Mathew slept until six o'clock a.m. on the days that he napped for two hours. On the days his nap extended to three hours, he was up at five o'clock a.m. You want your child to get enough sleep so that he's not overtired, but, at the same time, you do not want him to sleep so much during the day that he's sleeping short at night. If you need to awaken him, do so after ninety minutes to two hours into the nap. That way he'll have completed an entire sleep cycle and it will be easier for him to arouse.

When you awaken him, don't jar him from sleep or force him to get up. Instead, simply touch his shoulder lightly and say, "Snack is ready," or "We can go outside now." A gentle touch and an interesting invitation can help him arise more easily and in a better mood.

Add late-afternoon exercise According to sleep scientists, we typically fall asleep as our body temperature is falling toward a daily low and we wake up as it starts rising. The hormone cortisol drops off early in the night, then rises to high levels prior to our awakening in the morning. Exercise raises body temperature, so if you are trying to delay bedtime for your child, you may consider adding late-afternoon or even early-evening "exercise" to hold him off a bit. The key is to find something that alerts him enough to move his sleep time fifteen minutes but doesn't send him past his "window." A walk after dinner worked for Maggie's family.

"Maggie was our morning lark, but, as spring came, she was moving up to four-thirty a.m. We couldn't handle that, yet at night we couldn't keep her awake. She was asleep by six o'clock p.m. We found that if, after dinner, we took a stroll around the block, letting her walk or ride her bike, we could keep her up later. By doing so, we held her off until seven o'clock p.m., and she slept in until five-thirty or six o'clock."

Get the lights on If you have an early riser whose schedule you'd like to move later, turn on the lights in the evening. As dusk starts to fall, get the lights on and keep them on bright, until you are ready to begin the bedtime routine. You want your child's brain to be receiving the message that it's time to be awake.

Keep the television off in the morning. Early risers often have favorite shows that they awaken to watch. Reduce the incentive and the light by making a no-morning-television rule.

Move bedtime fifteen minutes later Delay bedtime fifteen minutes. I know you would like it to be more, but, at this point, if you attempt to shift more dramatically, it's likely that you'll end up with an overtired child who will awaken even earlier.

Delaying bedtime may be especially difficult when you are falling back from Daylight Savings Time. Not only is your child ready for sleep before the clock says it is bedtime, but so are you, making it more challenging for you to be patient and calm with your child. If you can, begin making adjustments for falling back from Daylight Savings Time a few days earlier, so that you can do so in small increments. Fortunately, it's actually easier to delay bedtime a bit than it is to make it earlier.

Wait I wish I could tell you that the next morning, after delaying bedtime fifteen minutes, a miracle will occur and your early riser will sleep in at least another fifteen to thirty minutes. Unfortunately, this is unlikely. It takes several days, sometimes a week—or longer, if your child really is a strong morning lark—for the body clock to begin to move even fifteen minutes. In the meantime, he'll probably still be awakening at his old time and, now short on sleep, be crankier than ever. Fortunately, positive results for your efforts should begin to show within three to four days. It is essential, while you are waiting for the shift to occur, that you not allow your child to pick up the fifteen minutes of sleep he missed by taking a longer nap. If necessary, wake him up after his normal amount of sleep.

It's also important to maintain the new schedule seven days a week to help your child shift. This is especially true if he's a strong morning lark, and if he's temperamentally irregular. Once his schedule has shifted fifteen minutes, begin the process again, moving back another fifteen minutes.

"Initially, when I moved Katrina's schedule, she started waking even earlier," Paul reported, "but I didn't quit. I was tempted, but I didn't. Instead, I worked with her, telling her it was too early. She needed to sleep until the clock said six o'clock. It took three weeks, but she got there."

"Kur" your child You can also help your child sleep longer by using the "kur" technique. When your child awakens too early, dampen a towel with cool water and lightly wipe his arms, legs, and torso, then send him back to bed. The result is a deep, restful sleep and more dreams. This technique is standard at European spas.

Use springing forward to Daylight Savings Time to your benefit When you turn the clock forward for Daylight Savings Time, what was six o'clock a.m. suddenly becomes seven o'clock a.m. You can choose to simply keep your child on his body schedule. Before the "switch," he went to bed at eight o'clock p.m. and arose at six o'clock a.m. Instead of keeping him on an eight o'clock-p.m. bedtime, leave him on his "body time." That means, in the spring, when the clock springs ahead, your child will begin to go to sleep at nine o'clock p.m. and sleep until seven o'clock a.m. If you're lucky, you may even be able to leave him on an eight o'clock-p.m. bedtime schedule and find that he will STILL sleep until seven o'clock, his normal "body time."

Go "cold turkey" Now that I've explained the "ideal" way to move your child to a later bedtime or wake time, let me tell you how many people do it, when they've been caught by surprise or didn't have the opportunity to take a gradual approach. Go "cold turkey." Simply move your child to the new schedule, changing wake, nap, and bedtimes as well as meals and exercise. Just remember, when you do this, that your child is out of sync and is going to need much more help winding down for sleep, because he's overtired by the time he gets there. And, despite the later bedtime, he will initially still awaken earlier than you'd like. Adjust plans and get your own sleep needs met as best you can, because your child is going to need more support from you to be successful.

Work together: teach new skills. Francie sighed. "We've tried all of those things. When we hold off five-year-old Elizabeth's bedtime, all we get is a crabby early riser. And anyway, at night, even if we're playing a game, or doing something fun, she'll wander off. I often think she's gone to the bathroom, but then I discover she's put herself to bed!"

If you, too, have an early riser, who seems permanently locked into celebrating sunrises, and she is three and a half or older, you can begin to teach her skills of what to do when she awakens early so that the other members of the family can continue to sleep. (If she's younger, you may have to accept the need for you to go to bed earlier and start your day earlier until she's old enough to work with you—sunrises really are beautiful!)

We approach this task not searching for a means to "make your child sleep," but instead thinking about the skills we need to teach to help her arise without awakening others. So, I turned to the group and asked, "What do you do when you awaken in the morning in order to avoid waking everyone else?" Their responses were immediate.

"I slip out of bed and go read in another room."

"If it's before six o'clock a.m., I usually just lie there and think about my day."

"I get up and go for a run."

What you realize is that, as adults, early risers have (hopefully) learned skills that respect their joy of early morning yet allow others to sleep. Little early risers need to learn those same skills.

Childproof the space If your child is going to practice the skills she needs to arise without disturbing others, safety becomes a critical issue. Cover electrical outlets and eliminate any electrical equipment. Initially, this may include lamps, CD players, and even a clock. Avoid potential falls by gating steps. Provide a sturdy stool so that your child can turn on an overhead light. If you live in a cold climate, dress your child in warm pajamas, so that when she's up early, she'll be comfortable.

Consider creating a "cubby," a cozy place for your child to go into and enjoy quiet play as the sun rises. "My husband opened the dormers in our daughter's room," Katie told us. "She loved to go in there and look at her books. When I got up, I often found the entire bookshelf emptied. I never minded, because she had let me sleep."

Provide toys Before your child goes to sleep for the night, together select toys that she can safely and quietly play with when she awakens in the morning. If you can, limit play with these toys to morning time so that they retain their "fresh" quality. Books are a wonderful option even for young toddlers.

If your child is not yet old enough to play on her own, or needs someone near, it is possible to support her yet still get sleep. "Sasha needed to be able to see us," Susan explained. "So, we put a mattress on the floor in her room. When she awoke early, I would go in and sleep on it. She could see me from her crib. That was all she needed. I could count on her to play in her crib for at least another twenty to thirty minutes while I slept a little bit more."

Include a snack Early risers have an appetite. Be sure you include a nutritious snack in your bedtime routine and leave a morning snack on your child's dresser. A half cup of dry cereal in a plastic container will often tide her over until you're ready to make breakfast. You can also leave juice or water in a cup with a lid, in case she is thirsty. Grinning, John reported back after leaving a snack for his early riser. "We added little marshmallows on Saturday and Sunday mornings, a little incentive to let Mom and Dad sleep in later on the weekends."

Clarify the rules Make sure the rules are clear. If you do not want your early bird to awaken siblings, be certain the limit is clearly stated. "You can get up and play, but you may not awaken your brother." Get your "backup" adult to vouch: "This is the rule!"

Teach to read the clock When you are comfortable putting a clock into your child's room, teach her to read it. If possible, use a clock with a face showing the entire dial, rather than a digital alarm. It's easier for a young child to get a sense of time when she can see the clock face. If you can't find one, or your child hasn't yet acquired the skill of reading it, set a visual clock when she awakens you. These are soundless timers that show a diminishing color as the time disappears. You can set it for up to an hour and get a little more sleep while your child watches for the color to disappear.

Use "picture planners" Francie couldn't wait to share her good news with the group. Elizabeth was still awakening, but she was letting them sleep. "What did you do?" I asked. "Star cards!" she exclaimed. I looked at her, a bit puzzled. "Picture planners," she explained. "Together, we made instruction cards with words and pictures, to tell Elizabeth what her 'tasks' were for the morning. For example, one card read:

Elizabeth

1. Stay in your room until the clock says 6:00 (with a picture of a clock showing 6:00)
2. Get dressed (with a picture depicting clothes)
3. Read books (with a picture of a book)
4. Don't wake your brother! (with a picture of a sleeping brother)

"We set the card next to her snack on the dresser. She loves to find it in the morning, and I firmly believe she is starting to 'read' from those cards. When she does all of the tasks on her card, she gets a gold star. After five stars, we bake cookies together."

Something tells me that one day Elizabeth is going to be a voracious reader and a great cook, with fond memories of time spent with her mom.

Think about your child. What skills can you teach to help him be more self-sufficient in the morning? Begin with realistic expectations. If he's able to play quietly for five or ten minutes, celebrate and encourage him by saying, "Soon you'll be able to make it fifteen or thirty!"

More quickly than you might ever expect, your early riser will hit puberty. Thanks to the hormonal changes that occur then, even your morning lark will be transformed into a temporary night owl. Instead of fighting to keep him in bed, you may find yourself dragging him out of it, wishing for the old days when you couldn't keep him there.

Temperament—customizing the plan to "fit" your child

It's starkly apparent, as you talk with others or observe closely the members of your family, that some individuals seem to adjust their schedules much more easily than others. Once again, the answer to why often lies in their temperament.

When your child is intense The challenge with intense children, especially extroverted intense children, is that they want company when they awaken and tend to bellow loudly when they don't get it. They move quickly from sleeping to awake and are ready to attack life with a passion.

Select "interactive" toys for them to play with in the morning. "Little people," stuffed animals, or building blocks and Legos may keep them occupied enough, "talking" and "creating," for you to get a little more sleep.

When your child is sensitive If it is possible, give this child the bedroom on the west or north side of the house and away from the street. Morning light and sounds awaken him. If the room faces the east or gets lots of light from streetlamps or other sources, cover his window with heavy blinds and take extra measures to stop the "leaks" on the edges. Use a fan or sound machine to block disruptive sounds.

When your child is slow to adapt Returning to sleep is a transition, a shift from one thing to another. Reducing stress and eliminating "alerting" sounds, smells, or changes prevents this child from awakening in the first place, and thus avoids the struggle to get him to "switch" back into sleep.

When you are a highly adaptable parent, your strength is your flexibility. But you may forget to teach your child that he "lives in a family." His future roommates and partner will appreciate it greatly if along with meeting his needs you also encourage him to learn the skills that respect others' need for sleep.

If you are slow to adapt, it is very challenging for you to awaken before your body and brain are ready. If your child is young and you do need to adapt to his needs, make the extra effort to get yourself to bed. It can also help if you take the steps, like putting up the darkening blinds and teaching your child skills, so that he will sleep and you can, too.

When your child's body rhythms are very predictable and regular If your child is a true morning lark, understand that you have all of his genetics playing against you. My father, a notorious morning lark, awoke at four-twenty a.m. to milk his cows until he was seventy-five years old. He proudly announced that he NEVER needed an alarm clock. In retirement, he was able to move his morning awakening to five-thirty a.m. That was it. You will want to teach this child self-sufficiency skills at a very young age. Adjusting his schedule more than an hour will be a very big challenge.

When your child is high-energy Like the intense child, the high-energy child is raring to go in the morning. Taking this child into bed with you only works occasionally. He tends to thrash and talk, more interested in playing than sleeping. Include with those morning toys something he can do to "move," like a rocking horse, little cars, or lightweight balls. If he's older, the perfect place to store a stationary bike may be in his room, so that he can pedal miles before you ever rise.

MOVING BED AND WAKE TIMES EARLIER

Our circadian rhythm runs on a twenty-five-hour cycle. As a result, it's easier to go to bed later and wake later than it is to go to bed early and wake early. That's why "springing forward" for Daylight Savings Time, or traveling east can be especially jarring to your child's system. Shifting to an earlier time requires your child to sleep and to awaken BEFORE his body says it's time. As a result, he can't fall asleep at night and wants to sleep in the morning, but because the rest of the world is on the new schedule, he can't, which leaves him cross and out of sync.

It had been ten days since we had switched to Daylight Savings Time—"sprung ahead" is the term commonly used. I asked the group what they were feeling like.

"I'm exhausted," Betsy sighed. Others nodded in full agreement.

"The extra light in the evening is energizing," Carla added.

Nicole said it was hard to get dinner eaten, because the kids wanted to be outside, playing. Jeremy had refused to eat until eight o'clock p.m. As a result, it was a real struggle to get him to bed. But he wasn't the only one. All of the kids were taking longer to fall asleep. Their naps were off, and they were waking up at odd times, their bodies not quite sure when they should be awake. They were also moody. Late afternoons were one continuous sibling snipe after another.

The reality is, we can predict this effect. It happens every year (unless you're lucky enough to live in states where they choose to ignore Daylight Savings Time). But if you do have the opportunity to "spring forward" every spring, you can expect challenging behaviors as a result of it, and plan for success.

Managing the tension

You can begin to manage the tension created by "springing ahead" by marking your calendar. Block off early mornings the first week after the switch—not just Monday, but the entire week. Often, Wednesday through Friday are tougher than the beginning of the week, because sleep deprivation has grown each day. By doing so, you eliminate the need to "rush" out of the house, adding more tension to an already challenging situation.

Eliminate evening activities, if you can, since you can predict the kids will be dragging. "By Wednesday, Nathan was exhausted," Karen explained. "I decided he should skip soccer practice, but I didn't tell him I'd already decided. Instead, I said to him, 'Let's talk about this. I think it would be better if we stayed home and got ready for bed earlier so you will have good energy for your tests tomorrow.' Amazingly, he agreed. Normally, he would have been upset, but there wasn't any fight. All he said was, 'I was looking forward to soccer tonight, but I know I'll go on Saturday.' " By allowing your whole family a bit of a break, you can make the move to an earlier wake and bedtime more smoothly.

And don't forget to plan for naptimes. If you can, take a nap yourself, because you can also predict that the kids will need more soothing, calming activities at night to help them "switch" into sleep.

Time: moving your child's body clock earlier

"I've got three kids, an eleven-, eight-, and three-year-old," Martha said. "By the end of summer vacation, they are all at least one hour to one and a half hours off a school schedule. I'm not even sure how to get them back on again." Whether it's that springing forward for Daylight Savings Time, or moving to an earlier wake time after a long, fun summer vacation, the changes need to begin in the morning.

Begin in the morning Begin by waking your child fifteen minutes earlier. Once again, moving in small increments makes it easier. Adjust meals and naptime as well, leaving bedtime on the old schedule. After two or three days, when your child is beginning to show signs of fatigue, move bedtime fifteen minutes earlier as well. Once your child is easily going to sleep at this new time, adjust the schedule another fifteen minutes, continuing until you reach the time you want. If your child is a night owl, make an extra effort to be consistent seven days a week, because you are going against his natural clock.

Use bath time to "trick" the brain Our brain slips into sleep as our body temperature drops. You can "trick" your child's body by bathing him ninety minutes before you want him to go to sleep. A bath raises body temperature. But when you get out of the bath, your body temperature falls, and, as a result, your brain is more likely to think it's time for sleep. Researchers have found this technique to be more effective for adult women than using pharmaceutical sleep aids, so feel free to try it for yourself as well.

Work together with your child "The other kids in the neighborhood were all outside, playing, and it was still light outside, but I knew that if I didn't get Michael back on a school schedule, he'd have a really tough start," Jenna told the group. "I started talking with him about the sun being different in the summer than it is in the winter. He's entering kindergarten and really didn't understand this. I had to explain that it was still the same time, even though the sun was still up and some kids in the neighborhood were outside playing. Then, to make it more fun, I decided to use our Visuals for Learning picture-planner board. We had used it before. He knows the schedule, but it was fun for him to lay it out. 'What do we do first?' I asked him. He picked the picture for snack. Then I asked him, 'What do we do second?' He chose the picture of pajamas. We continued until we had our entire routine. I let him play with the magnets and put them on the board where he wanted them. I only did it for a week. Now it's sitting on his dresser, where he can see it. It helped us refocus on the routine instead of on the kids playing outside and the daylight."

Think about what your child needs to know. It's very likely that he does not realize that school will be starting in a few weeks, or that Daylight Savings Time will be occurring. You can win his cooperation by pointing it out and working with him.

Older children will benefit from a gentle reminder and a "curfew" that helps them to begin to ease back into a schedule that will allow them the sleep they need. Talk about the importance of sleep with them. Ask them questions, like "If you choose this movie or activity, how will it affect your bedtime?" Help them to recognize that losing sleep will impact how they feel and how they perform.

Add exercise Children who are calm and physically fatigued fall asleep more easily, even if the time is earlier than normal for their body clock. Add exercise to your child's schedule to help him make the change. Martha was delighted to report, "Our eleven-year-old didn't appreciate it when, three weeks before school opened, we began to clamp down on her bedtime. But we added lots of swimming and bicycling to her day and, despite her protests, she couldn't stay awake." Tired children fall asleep earlier and easier.

Go "cold turkey" Adjusting to an earlier bedtime and wake time can easily take three to four weeks for full "recovery"—unless your child is exhausted. When this is true, I often find that it is possible to simply move the child's bedtime an hour or even two earlier. By doing so, you "hit" his real window for sleep, and he falls asleep easily and sleeps all night.

Seven-year-old Maggie had been staying up until at least ten p.m. when I met her. After I reviewed her sleep journal, it became apparent that she was only getting eight hours of sleep. But even when her parents tried an earlier bedtime, she didn't fall asleep. "We had tried putting her down at nine o'clock," her parents explained to me, but she was still up until ten o'clock or later. Initially, when I suggested an eight o'clock p.m. bedtime, they hesitated. "We knew it would never work," they told me later. "But then we talked about it and decided we could at least try it. You can't imagine how amazed we were when she actually went to sleep."

When Maggie's parents had put her down at nine o'clock, she was already past her "window" and well into a second wind, but eight o'clock caught her natural "window" and let her fall asleep more easily. So, when you need to move your child's bedtime with very short notice—if you just didn't have the opportunity to plan ahead, or you realize your child is very tired—you can try going "cold turkey" to the new bedtime.

TEMPERAMENT—CUSTOMIZING THE PLAN TO FIT YOUR CHILD

Just as it is true for waking early, each child's individual differences will matter when it comes to changing bedtimes. It's the slow-to-adapt, regular night owl who will be the toughest to move to an earlier bedtime. The sensitive child may find that falling asleep when it's light is very difficult. He'll need you to take those extra measures to darken his room so that his brain knows it's time for sleep.

Celebrate success

Waking up eager to start the day requires waking in sync with your child's and your own circadian rhythm. Sometimes, achieving just the right fit requires changing your schedule. It's a process that requires thought and patience. Once you've made the changes, cheer for yourself and for your child. You worked together and did it! Celebrate your own patience and ingenuity and your child's growing awareness of how important sleep is to his well-being.

SEVENTEEN

TRAVEL AND HOLIDAYS

Planning for success

"The good thing about travel is that it takes you to new and different places. The bad thing about travel is that it takes you to new and different places."

—Diane, mother of two

Becky awakened at three-thirty a.m. in order to get herself and her three sons ready for a visit to her parents who lived 1,500 miles away. By four-thirty a.m., she had the boys up, dressed, and ready to go. Her hope was that they'd go back to sleep on the way to the airport or on the plane. Of course, they did not. The baby slept thirty minutes instead of his usual two hours. The preschooler didn't nap at all. By the time they arrived, everyone was cranky and tired, and at bedtime, they were so overtired they couldn't settle. "They wanted me to play with them, but I was too tired to play," she told me. "I was impatient with them, when normally I'm pretty good, and I got irritated when they couldn't fall asleep. I knew everything was new, but it overwhelmed me, because I was so tired, too."

Like Becky, thousands of other parents have shared their travel horror stories with me. I, too, have faced my own challenges traveling with children. I have always loved to travel, so, when my son was born, I thought my first book might be about traveling with children. After a few disastrous trips with him, I realized I couldn't write the book, because it would be one-sentence long: Stay home! The reality is, however, that we don't always, nor do we wish to stay home. Learning how to travel well is an essential life skill for our children to learn.

Today, my son is an adult with a degree in international business and has lived and traveled all over the world. So, I can tell you from personal experience, there is hope. What I discovered in the process of helping my son to be successful is that learning how to adapt to the resulting time changes and potential disruptions in routine is essential. You can begin by managing the tension.

MANAGING THE TENSION

Sleep deprivation is a source of tension. If you begin your trip tired or on a schedule that immediately puts you out of sync with your circadian rhythm, you are inadvertently increasing the odds that the first few days of your trip will be filled with more power struggles than fun.

Begin with travel plans

An enjoyable trip or holiday celebration begins with thinking about what each member of your family needs, to be successful. In order to make the travel plans that fit your family best, think about the effect of those plans on the body clock.

I couldn't miss it. Paul was giving me the "look." "I don't want to stop and think," he said. "I want a break. That's why I'm going." Catching myself again, I turned to the group. "What do you think about when you travel or plan holiday celebrations with your kids?" They jumped into the conversation.

"I just want someplace warm."

"I miss my family and I get so excited thinking about spending time with them."

"I want a break in the routine."

"I dread holidays."

It's true, it's not easy. And thinking or planning ahead is unlikely to be at the top of your "to do" list. But whether you're looking for sun and fun, a good time with relatives, or "surviving" time with relatives, the kids will be more flexible, you'll all enjoy yourselves more, and you'll be less likely to "pay" afterwards if you stop and think—even just a little.

Early and late flights tend to be cheaper, but there is a cost to them in the disruption they create. Sometimes it's worth it and sometimes it's not. As you begin making your plans, ask yourself, do you have an entire contingent of night owls? If so, a later flight or drive time may work well for you. You want to avoid driving during your normal adult bedtime, however. That's when you are most vulnerable to falling asleep at the wheel.

If everyone is up and raring to go in the morning, an early schedule may fit you best. A "mix" of types may require you to consider who seems to recover or work out of type most easily.

"I just chose the cheapest flight," Sarah wearily told the group. "But we ended up arriving late eastern time, so it was 'bedtime.' But the kids were not ready for bed, and we were all in the same room. I tried to get them down because the next morning we were supposed to be up and out early for an outing. It made for a very tough start."

If you do need to choose the flight that disrupts your child's schedule, plan for "entry" time at your destination. Instead of hitting the hot spots immediately, consider a leisurely time at home with the relatives instead of rushing out. If you are staying at a hotel, plan time to simply mellow out by the pool for the first day. Taking time to allow your body clock to catch up with you a bit will let you enjoy the rest of the trip more.

By making plans that fit your family, you immediately reduce the tension associated with travel. You begin your trip with smiles instead of screams.

Create travel rituals

Travel is "alerting," which can be disruptive to sleep. One way to enjoy the opportunity and still ensure that everyone gets the sleep they need is to bring along the "lovies" and cues that are familiar. That's what Sharon discovered. "When we are away, the kids want us to use the same routine as we do at home, so I have to try very hard not to forget the books!"

By creating travel "rituals," you soothe and calm your child. How do you do it? Bring along his favorite blanket, stuffed animal or pillow. If the pillow is too bulky, bring the pillowcase, so that the strange pillow smells right. Take along a small CD player with your child's normal bedtime music, or make a recording of the typical night sounds in your home, then play it when your child goes to sleep. If you don't smoke, make sure that you get a nonsmoking room away from the smoking area. Any hint of smoke triggers the brain to be alert. It's part of our survival instinct and is especially strong for highly sensitive children. Take along a night-light and comfy pajamas. This is not a time to introduce brand-new ones.

Plan for what you'll do when lights are out

It's very likely that you will all be sleeping in the same room while you're on vacation or visiting relatives. Instead of being frustrated that the lights are out and you're in a hotel room with the kids, bring along a headlight (it straps to your head, leaving your hands free) so that you can put the kids to bed and read. If you like, have a glass of wine. Or ask for a room with a patio or deck where you can enjoy watching the stars or the lights of the city while the kids sleep. And don't forget to give yourself permission to go to sleep and get the rest you need.

If you are visiting relatives, plan ahead whether you'll put the kids to bed together with your partner, or if one of you will take care of it while the other visits. It's easier to decide at nine o'clock a.m. than it is at nine o'clock p.m., when everyone is already overtired.

Monitor the stress of watching small children in new places

I had met a friend and her two young children in a park along the Mississippi River. The river tour boat goes out from there, and, on a whim, we decided to take it. Wild with excitement, two-year-old Cameron immediately started running up and down the deck of the boat, thoroughly enjoying hearing himself screech. But a fellow passenger did not think it was so delightful. Cameron needed to be redirected. Telling him not to screech wasn't all that effective, it was just too much fun and he was too excited to stop. So, I asked him if he could walk backwards from the front of the deck to the back. When you are two, walking backwards is not an easy task. In fact, it takes a great deal of concentration and effort. It focused his energy for at least fifteen minutes, but the tour was ninety minutes long. We had to come up with more ideas. Watching Cameron's glee with this new experience was fun, but by the time the boat docked, both Mom and I had worked very hard and were ready for a break!

It reminded me once again of the importance of making plans and then cutting them in half, recognizing that not only do you need energy to cope with your own reaction to the new situation, but the stamina to be able to help your child be successful as well. Be aware, too, that you may be more comfortable letting the kids do things that grandparents find "alarming." Everyone will feel their tension, so consider this and work through it.

Remember also that every home is a new opportunity for exploration by a young toddler, and families without a young child may have forgotten the importance of "childproofing." Instead of visiting everyone, invite them to come to you—that way, your child can become accustomed to one environment, you can childproof it, and everyone will be calmer.

By thinking ahead, you'll ensure that you have the energy to have fun rather than grit your teeth, frustrated with your child and anxious for the event to be finished or the visit to end.

Allow time for reentry back home

You can predict it: no matter how well you have organized your trip, or how restful and wonderful it's been, it's still a shock to return home to everything that has been piling up while you were gone. Post-vacation or holiday blues is real. Plan for it.

If you return on Sunday night and have to be up the next morning for work and school, you can predict a week of power struggles. Why end your trip tense? Shorten your venue by a day—maybe even two—and relax at home before you have to go back out and hop on the train of life. Allow yourself to unpack and unwind, buy groceries and get back into the routine. That way you can actually have time to enjoy fond memories of your great trip.

When that's just not possible, plan for an "extra energy" high-protein breakfast the morning of your reentry. After returning late Sunday afternoon from a weeklong trip with his family, and having to get up early Monday for school, eight-year-old Mathew exclaimed, "Mom, you were right about the extra energy from breakfast. It really helped me get through the day!"

TIME: ADJUSTING YOUR CHILD'S BODY CLOCK

The more you can stay in sync with your child's body clock, the easier it will be for him to cope with the change in surroundings. A disrupted routine upsets the body clock, making it more difficult to fall asleep, and reduces the amount of deep restorative sleep your child will experience.

Maintain your routine

It was a wedding that brought the family together from across the nation. Friday night, the cousins all swam in the pool, but afterwards, instead of socializing, Evan and Deb disappeared with their kids. "We knew our kids needed sleep," Deb told me later. "Troy, our little one, had only napped thirty minutes during the five-hour drive there, instead of his usual two hours. Unfortunately, he didn't sleep well that night. We suspected he was overtired. So, the next afternoon, instead of taking him to the wedding, Evan stayed with him at the hotel and let him nap. We all attended the reception that night and had a great time. If we hadn't let him have his nap, he would have been a bear, and no one would have enjoyed themselves."

If your trip is short and the time change minimal, keep your family on "home time," and maintain your regular routine. Throwing off your child's sleep patterns ends up causing sleepless nights.

If the time change is drastic, or it's not possible to do so, follow a pattern similar to the one you do at home, spacing meals, naps, wake times, and bedtimes accordingly but on the new time frame. If possible, even start the routine a few days before your departure, going to bed earlier, if you're traveling east, and later, if you're going west. That way, your child's body will have begun the adjustment in familiar surroundings.

In your new environment, get outside in the morning as much as possible in order to reset your body clock.

Honor naps and siesta time

"We were all having such a good time, and Emily is very flexible," Deborah told the group wearily. "So, when our family met for a vacation, we entirely let our schedule go, skipping naps, going to bed late, eating at odd times. Honestly, she did fine while we were there. I think the excitement and all of the attention from the relatives kept her going, but you can only run on adrenaline so long, and then you crash. It's been nearly three weeks since we returned, and we're still 'paying' for it with disrupted nighttime sleep and a very crabby little girl."

Stopping to rest often leaves you feeling as though you're "missing out," but it's really a matter of quality versus quantity. When you plan a break in your day, you will enjoy your experiences more, because everyone will be well rested. If you don't, you may see or do more, but it's likely that your memories will not be of the event but rather of the meltdown that occurred while you were there.

John paused thoughtfully. "It's not easy for me. When I have paid for tickets, I want to maximize my time. Anyway, we had tickets for Disney. We hit the park first thing in the morning, when the crowds were lighter. By eleven-thirty, the kids were ready for a break. I was, too, but I never would have stopped. My wife convinced me it really was worth it to go back to our hotel for lunch and siesta. We did. Afterwards, we swam in the pool, then returned to the park later, when the crowds had thinned out. I hate to admit it, but it was the most enjoyable vacation we've ever had."

Kathy waved her hand wildly, eager to jump into the conversation. "Over New Year's, we went to Disney, too. We did a similar thing, arriving early at the park and taking time to rest. We had been talking about the fireworks and New Year celebration. The morning of New Year's Eve, our seven-year-old woke up with a very serious look on his face. 'Mom,' he said. 'We're going to want to stay up late tonight. Maybe it would be best if we hung out by the pool and had a quiet day. Then we'll have the energy for tonight.' I couldn't believe it! But that's what we did and it worked!"

"Most people don't realize how much their daily activities can affect their nightly sleep, especially when on vacation," Richard L. Gelula, executive director of the National Sleep Foundation, says. While on vacation, we tend to overschedule ourselves, eat more, snack more frequently, and load up on chocolate and caffeine. Like my grandmother said, naps are good for relationships. They also greatly increase the odds that vacation will be restful and *fun,* because your child won't be overtired and unable to fall asleep in the new setting.

Don't forget the relatives

"It's not naps that are the issue when we visit my in-laws," Michele groaned. "It's the mornings. My father-in-law likes to sleep in, and we have early birds. If they wake him up, he's impatient and grouchy all day long."

It's not just the kids who cope well when they are in sync with their circadian rhythm, it's the adults, too. Once again, you can plan for success. If Grandpa likes to sleep in, plan to get up quietly and go out for breakfast, or, if the climate is nice, put together a breakfast "picnic" and head to the backyard or the local park. The kids will get exercise, Grandpa will sleep, and you won't be walking on eggshells trying to keep the kids quiet. Your visit will go more smoothly.

Michele agreed to try the plan the next time they visited the in-laws. A few weeks later, she was grinning as she arrived at the family center. "We made the six-hour drive to the in-laws. On the way, we created a secret plan, to sneak out of the house and go out for breakfast. The kids really bought into it. They were wonderful. After breakfast, we stopped in the park. By the time Grandpa awoke, they'd eaten and played outside. They were so much calmer, and so was everyone else."

When our schedules are upset by travel or family visits, our bodies don't know when to be alert and when to be asleep. Falling asleep becomes more difficult. You or your child may find yourself lying there, vibrating with energy because your brain doesn't know when to shut down. It's the erratic schedule that is most disruptive to the body clock. Crazy hours make crazy kids and adults, so think about your routine and, to the best of your ability, respect it—for everyone.

Plan exercise

The nice thing about exercise is that it not only helps to set the body clock, but it also reduces tension. Often, however, it's the first thing to be dropped from the schedule if you are driving a significant distance, or flying. Make a conscious effort to include it on the agenda.

If you're driving, plan to stop every ninety minutes to two hours and let everyone out to move and get a breath of fresh air. Bring along a ball or a Frisbee. Look for a grassy open space and let everyone run! And don't forget a cooler. You can predict that, after exercising, everyone will be hungry and thirsty. If you have grapes, water, bananas, and other nutritious foods available, you'll avoid snacking on caffeinated drinks, candy, or other sweets. That way, you won't inadvertently diminish the positive benefits of exercise by increasing blood sugar and caffeine levels.

Not only are these breaks good for your children, they are also good for you. The National Sleep Foundation reports that 100,000 automobile crashes, 71,000 injuries, and 1,550 deaths a year are the result of drowsy driving.

When you can't move big muscles, move little ones. Bring along stress balls, wikki sticks, rubber bands, finger puppets, or other "finger" toys that let everyone fidget.

Beware of over-the-counter drugs

Unfortunately, one of the perils of travel is often illness. Changes in water, recycled air on airplanes, and disruption of sleep all tend to make us more vulnerable to illnesses. Be aware that many over-the-counter drugs cause sleeplessness at night. Triprolidine is the active ingredient found in Actifed and pseudoephedrine, a nasal decongestant found in many over-the-counter cold remedies. All of these ingredients can keep your child awake at night and sleepy during the day.

Drink water

It's easy to become dehydrated while traveling, which not only leaves your child feeling unwell but also awakens him in the middle of the night, desperately in need of a drink of water. If he's highly sensitive, the odds are that the temperature and taste of the local tap water will not be acceptable, leaving you with an upset child at two o'clock a.m., as you dash to the vending machine for bottled water.

Pack water bottles and a soft-side cooler to ensure that your family gets the liquids they need. Water is the best. Sodas are often diuretics, leading to more frequent urination. Rather than quenching thirst, they actually increase it as fluids are eliminated by the body.

Work together

Ultimately, you want your child to take over planning for his own needs as he travels, so share your concerns and considerations. Begin by talking about the trip. Make a picture planner if he's not reading yet, or jot down a clear list of expectations if he is. Let him know where you will be going and where you will be sleeping. What day will you be visiting relatives? When will you be going to the amusement park? Will there be time for the pool? Teach him words like "May we please go back to the hotel now, I'm getting very tired." Or "I'd like a break." Or "I'm having trouble settling, would you please massage my back?" By giving him words that are respectful you help him to express his needs appropriately. It's much easier to be empathetic and sensitive to someone who is using words instead of sulking or throwing a tantrum. Remind the adults to use them, too!

Lydia laughed. "We've talked about 'bubbling inside' with the kids, but my husband always insisted that he never experienced bubbles. After four hours at an amusement park, I saw this sheepish look on his face. When I asked him about it, he admitted, 'My bubbles are up. Let's take a break.' We all laughed. The kids loved it that Dad was 'bubbling.' "

TEMPERAMENT: CUSTOMIZING THE PLAN TO FIT YOUR CHILD

When your child is intense

It's not just your child's body clock that needs to be considered but his temperament as well. If you know your child is intense, you can predict that, upon arrival, he'll be excited and wild with energy. Plan to arrive early in the day, so that he has time to work through his excitement before bedtime.

"We choose the red-eye when we go to New York from California where we visit relatives," Sam explained. "They don't get much sleep, but they are so excited that they can make it through the entire day, and then we put them to bed on eastern time. We always plan our trip the week before school starts, so, by the time we get back to California, they are off their summer schedule and ready for an early school start." By working with your child's intensity, you can help him be successful.

When your child is sensitive

"I'm too hot." "I'm cold." "My pillow smells funny." "I can hear the television next door." "There's someone in the hallway." It's the sensitive child who has difficulty settling, because of the sensations that bombard him in a new environment.

Check out the room you'll be staying in. Is it near the elevator or ice and vending machines? Your highly sensitive child is likely to hear every "clunk." Ask for a room away from busy streets and other intrusive noises. Bring along a familiar air freshener, to mask unfamiliar odors. Consider shopping in local grocery stores and "picnicking," thus allowing your child to eat familiar foods. Drink bottled water, even if you're simply visiting another part of the country. Highly sensitive children often experience diarrhea and upset stomach from changes in water. It's tough to sleep when your tummy hurts.

On the plane or in a car, give your sensitive child mint-flavored gum to mask some of the smells. Bring along headsets, so that everyone can control their audio space. And, despite the fact that you may have invested significantly in the portable DVD player, use it sparingly. The combination of speeding along, the flash of light, and the movement on the screen can leave the highly sensitive child completely overstimulated by the time you arrive.

And don't forget to pack the sunscreen. This child is very sensitive to the pain of sunburn. Help him to feel comfortable enough to sleep by ensuring that he doesn't get burned.

When your child is slow to adapt

The slow-to-adapt child immediately moves into a state of alert in new places or when shifting from one thing to another. It's this child, more than any other, who needs his "trip planner," continuation of his routine, favorite "lovies," and familiar foods. He travels best when you establish a "home base" and move out from it, so that he can become comfortable with his "new nest" and rest well. You may find that instead of getting his own bedroom at Grandma's, he'll sleep best if you let him camp out on an air mattress in your room. That way, he will have the security of being near you, but you will still each have your own space. It will also be easier, when you return home, for him to go back to his own bed than if he slept with you.

If you are quick to adapt and irregular yourself, it's easy to forget how important it is for your child to maintain his routine. He'll be more successful, and you'll all have more fun, if you help him out by being consistent.

When your child is high-energy

Going on vacation with my husband is often like going to boot camp. He arises in the morning, eager for a run or bicycle ride. Then it's time for swimming or water skiing. And then again, a leisurely hike may be what comes to mind. Lying quietly, reading, or simply enjoying the warmth of the sun lasts about fifteen minutes. This is a middle-aged man!

If you have a high-energy child, he needs to move. If you enjoy quieter activities, enlist the help of others to ensure your child gets the exercise he needs to be ready for sleep at night, otherwise he'll be showing you the "spring" in the mattress of every new bed. Select a hotel with a pool, and use it. Include walks, bicycle rides, stops at local parks, and any other physical outlets you can find. Remember, too, that standing in line is going to be tough for this child.

High-energy kids will also keep going, so it's easy to think you don't need to maintain their routine, but they, too, will crash.

When your child's body rhythms are very predictable and regular

Regular kids tend to fall asleep anywhere, but just as their body clock is set for sleep, it's also set for meals. Be sure to carry a soft-sided cooler and stock it with nutritious foods. When it's lunchtime at home, this child is going to need something to eat!

When your child's body rhythms are unpredictable and irregular

Irregular kids go with the flow, which is great for travel, except they, too, become overtired. It's tempting to skip or disrupt their routine, because they are so flexible, but in the end, they, too, will benefit from consistency.

Celebrate your successes

Travelling with children can create wonderful memories for your family. All it takes to help them be successful is a little thought about how you'll all get your sleep. "I persuaded my husband to take a more low-key vacation this year," Emma said. "We booked a short flight, no time change, no arriving after bedtime, and no changing hotels. Initially, he was reluctant, but I have to admit, it was so relaxing. It was just what we all needed."

Plan for adventure, expect fun, and take time to sleep.

EIGHTEEN

CHANGING BEDS

Moving out of your bed, out of the crib, or from one bed to another

"I got him into my bed, now how do I get him out?"

—Dana, mother of one

Imagine for a moment that I am taking you on a trip to the Boundary Waters Wilderness area. You must pack everything you'll need for the trip, because there are no stores there and, since your cell phone is unlikely to work, no deliveries, either. Everything you pack must fit into a canoe, and when it's time to portage from one lake to another, you'll have to carry all of it on your back, as well as your canoe, paddles, and life jacket. There are no cars, trailers, or roads, for that matter, only water and rough walking paths strewn with rocks and roots jutting up to catch your foot. At night, the stars may shine brilliantly, and, if you're lucky, Mother Nature will display her fireworks, the Northern lights. This light, along with your campfire and flashlight, will be the only illumination available once the sun goes down. You'll sleep in a tent that you'll crawl into after hanging your food in a tree so the bears won't get it. The latrine is about three hundred feet down the path, through the woods. And if there's a storm, you'll take whatever Mother Nature throws at you, since there is no place to go for shelter.

Now, envision that when we arrive, I surprise you by telling you that I'm leaving you to manage on your own. I paddle away. You do not know where I've gone or if I will return. Are any feelings of panic arising? How well do you think you'd sleep?

Now try another scenario. This time, we go together, and I stay. Perhaps we even share the same tent, or maybe we set up side by side. We talk about the fact that some day you may choose to do this on your own. You ask questions, striving to learn what you'll need to know. I show you, work with you, and celebrate your successes with you. We agree that the first time you "solo," I'll camp within earshot of you, right down the bay. Or, perhaps, you may decide that rather than going "solo," you'll bring along a friend and share the experience. During the entire process, you will see the vision, understand the plan, learn the skills, and know that someone you trust will always be available to support you as you practice. Then you'll do it!

Your child knows as much about sleeping in a new bed or down the hall as you may know about wilderness camping. And, if you are a wilderness camper, you realize that you have learned your survival skills from a supportive mentor and through much practice.

Just as we would insist that our child approach his first wilderness trip gradually and with a trusted guide, so, too, do we want to move toward helping him learn to sleep in a new bed, or down the hall. It's important that his first experience be a positive one. There's no need for "abandonment" on his first "solo." We can help him be successful by approaching the process gradually and working with him.

Know where you are going

It's important to think about the changes you are about to make. Are you switching whom your child sleeps with? Are you shifting him from one bed to another? Is the room where he'll be sleeping changing as well? Perhaps the change involves all three.

By clearly defining where you are going, you can begin to break the process down into steps. A child who has never slept alone has to learn about that experience, before he's ready to explore sleeping in a new room as well. A child who has been sleeping in a crib needs to understand the rules and the sensations of a "big bed" first. And a child who is shifting rooms has to feel comfortable in the new setting before he can sleep soundly there.

While the changes may vary slightly, by following a few basic steps, you can help your child make the move smoothly and successfully. The steps include:

1. The decision to "move"
2. Building a new "nest"

3. Learning to sleep independently

4. Supportive practice

Changing a sleep site creates tension. As you review the steps, you will realize that they focus on managing the tension. By doing so, you help him to approach the alterations calmly rather than be pushed into a state of alert by them. The process begins with you.

The decision to change begins with you

Even if your child initiates the process of change by climbing out of his crib or asking to change his sleep site, you still have to decide how to respond. You have to be prepared for the "event," believing that your child is capable and ready for the challenge—and that you are, too!

Laurie and Marc took their lead from their son. "Our son is eight now, and still sleeping in our room. It's not as though we have never tried to move him. We have. He has slept in his own room successfully, but the pattern has only lasted a few weeks, then he always comes back to us. Any disruption in his routine starts the cycle. If we have company, or leave town, even for a mere weekend, we are back to square one. In the past, we haven't minded too much, but recently, Matt came home and showed us a picture that he'd drawn of himself sleeping in his own bed. He didn't want the other kids to know he still slept with us. Immediately, we realized it was in his best interest—and ours—for him to move. He wasn't getting good sleep in our room, and it was deterring him from accepting overnight invitations with friends as well. So, we sat down with him to talk about making the change. We were very serious. 'It's time,' we told him. Somehow, he recognized that this time was different."

Just as your child has to be ready for the change, you need to be, as well. Rebecca intuitively grasped this idea. "My in-laws were adamant that Libby should be moved into her own room when she was two months old. They were convinced that we were starting a very bad habit. But my husband and I just weren't ready. We wanted her in the bassinet next to us. Around three months, we started putting her down for her nap in the crib in the nursery, and around four months, we began to realize that she liked sleeping there. Then we were ready, and so was she."

If your child has loved the safety and security of his crib or bed, it's unlikely that he will eagerly let go of it. And if you have shared your bed with your child, snuggling together has communicated a special kind of "I love you" that is difficult to give up. As a result, your child—not surprisingly—treasures the sense of security he has found in his existing sleep site. So, if you decide to change or adjust it, first ask yourself if you WANT to do it or if you're feeling that you SHOULD. If the motivation for change is because others are telling you that your child is too old to be sleeping with you—or in his crib, or with his sibling—choose to wait. If you are driven by the arrival of a new baby, or an impending move, but don't believe your child is quite ready for a change, find a different solution.

Unless you are thoroughly convinced that making the move is in the best interest of your child and your family, your child will sense your trepidation and find it more difficult to settle into his new sleep site. You also won't have the energy or the commitment to support him as he makes the change.

But you may find that the little hand in your face no longer feels like an "I love you" and instead is edging toward being intrusive and uncomfortable. Or your child's crib has become confining rather than comforting to him. Or he appears to be self-conscious or embarrassed by the present arrangement. If these or any similar situations arise, then it's time for a change.

If you are living with a partner, take time to talk, so that you can work together. Decide what is best for your family and how each of you will help your child. Listen carefully to one another, striving to understand each person's point of view. Avoid negating fears, or apprehension. Brainstorm potential approaches until you find one that feels right for your family. Working together will allow you to remain calmer. Your child will sense your confidence and know that it is time to begin to practice. Don't worry, we'll help her get there without leaving her to cry and disrupt everyone else as well.

Find the "window" for change

It's less problematic to change your child's sleep site if you pick the right time. By waiting and watching, you'll discover that there are "windows" when the move can be much simpler.

It didn't surprise me when Paul volunteered, "All of our babies slept with us the first three months, and then we started laying them in a crib next to our bed. From there, we moved the crib into a nursery. It was very easy to do."

Four months of age is often a perfect "window" for shifting your child from one sleeping arrangement to another, whether that is from your bed to a crib or from a cradle to a crib. At four months, there is a significant change in your child's sleep patterns. He's through the initial upheaval of birth, has learned more about his new world, and is often ready to take on the challenge of sleeping independently.

But if, at four months, you're not ready, or your child vehemently lets you know that he's not prepared to shift, don't worry. There are more "windows" to come. Instead of aiming for the typical markers of nine, twelve, or fifteen months, or other points of significant growth, select a time in between them. When you choose a point where your child is in a state of equilibrium rather than in the midst of a significant developmental upheaval, he has more energy and reserve to manage the challenge of a change. Every child is going to be slightly different, so watch your child closely. He'll provide you the clues that he's ready for a change.

Watch and listen to your child

Look for a point right after a major growth spurt. Note when he has learned a new skill or has experienced a spurt of development such as cutting new teeth, crawling out of his crib, voicing an explosion of new words, learning to read, or a new awareness of his own body. Then you will know the particular growth spurt is complete, and he now has the energy and stamina to face a new challenge.

If you can, avoid changing your child's bed in conjunction with the arrival of a new baby. This event is already stressful. Instead, make the change months before or months after the baby's arrival, even if it means borrowing an extra crib from a friend or relative.

When you move, make setting up your child's bed and sleeping space a top priority. Rather than introducing a new bed with the move, you can reduce the distress of the upheaval by putting up the "old" bed in the "new room." Later, when he's comfortable with his new space, you can choose to change his bed. Otherwise, the combination of moving *and* changing beds and bedrooms may be more than your child can handle, leaving him much too upset to sleep well.

While there are times to avoid changing your child's sleep site, there really isn't one perfect time. You can allow your child to lead, or, if he is a bit slow to adapt, you may choose to provide a little encouragement. Together, you can decide when the time is right for your family.

When the change is abrupt

Sometimes, you get caught off guard. The house sells before you expected, the new baby arrives early, or some other event leaves you scrambling to make a quick change in your child's sleep site. That's what happened to Bob. "Our home sold the first day on the market, and the new buyers needed it at the end of the month. The only way we could make it work was to move in with my parents. Charlie loved his crib. He's the kind of kid who would still be in it at four, if we hadn't moved. Unfortunately, there wasn't any place for his crib at my parents. So, we moved him in with his sister. He's a pretty mellow kid and, thankfully, handled it just fine."

But if your child does not have a low-key personality and there's no sibling to ease the stress, it's also important to know when to "revert" to your child's old sleep site. If the change is too abrupt for your child, recognize it. He'll let you know with his vehement protests. If this is the case, find some place, any place, to get his crib back up, or invite him back to your bed, or to his old bed—for a while. It will be time to practice again when things have settled down.

Share the vision

If your child is a year or older, begin talking about where you eventually want him to sleep. If you can, set up the new bed in his present sleep site. If there isn't room, set it up somewhere else in your home, preferably in the room he'll be moving to. You can choose to start simply by saying, "One day, you'll be ready to take your naps there." Or you may prod a little more by adding, "Pretty soon, you'll be ready to sleep *all night* in the big bed." You can even invite him to practice by asking him, "Is this the night you want to sleep in your big bed?"

When he says no, you can say, "All right, but pretty soon you'll be ready."

Nancy used this strategy. "We put up the bunk beds in Stacy's room and then talked about them for six months. She was excited and wanted to sleep in them. By the time we let her do it, she slept all night long."

Keep your vision positive. Laughing, Martha said, "We let him know that his cousins slept in a big bed. Every night, we'd go through the list. Rachel sleeps in a big bed and Michael does, too. He'd get this very serious look on his face, and ask, 'Does Ryan sleep in a big bed and Joanna, too?' We'd nod and tell him, 'Yes, they all sleep in big beds and one day you will, too.' It was so cute; he even wanted to know if the mailman had a big bed as well."

John agreed. "Lindsey got excited when we told her that she'd be able to spread out over the entire mattress and get *all* of the covers."

Even Noah, who had slept with his parents for his first two years, was ready for the change when his parents created a positive vision for him. "We didn't want it to be a 'cold turkey' move," they said. "Instead, we put up a bed for him and explained that it had been his daddy's bed when he was a little boy, and now it was Noah's. He practiced taking naps in it and spent time playing there with us. It became a good place to be. Soon, he wanted to sleep in it, just as his father had done. Once he was comfortable sleeping in it, we moved it to his own room."

Change begins with a vision. A picture of what's going to happen. It's not a shove, nor a command, but rather a gentle guide, pointing the direction your family chooses to go.

Create a new "nest"

Whether your child is moving from one room to another, your bed to his own, a crib to a "big bed," or simply from one bed to another, you are altering his "nest." He has been comfortable in his old sleep site and safe enough to sleep there, so, when you change it, it's important to make the effort to create a new "nest," where he also will feel safe enough to sleep.

"That must be what's happening!" Lydia exclaimed. "The baby is due in a few months, and we needed to move Eric out of the crib and the nursery before her arrival. We didn't want it to be traumatic for him, so we really thought about how to make the transition as smooth as possible. We bought him a new, big-boy bed and prepared the room. He helped us paint it and even carried all of his stuffed animals into it. We read his books and played there.

"Saturday was chosen as the 'transition' night, because both my husband and I would be home. The first night we went through our normal routine, he climbed right into bed and fell asleep. That lasted three nights, but on the fourth night, he started screaming the minute we said time for bedtime snack. He was adamant. He was NOT going upstairs. It affected his naptimes, too. He started crying at lunch, knowing that nap immediately followed. I thought it must be a phase or something. I couldn't understand what was happening. But now that I think about it, he had said, 'Eric sleep in old bed.' Changing rooms has been our focus, we never thought about the fact that he might be missing his crib."

That night, Lydia and her husband put Eric's crib up in his new room. Eric happily crawled into it and went to sleep. The pre-nap and bedtime tantrums miraculously disappeared. Each week, they asked him, "Would you like to try sleeping in your big-boy bed?" Eric refused until four months later. Then, on his second birthday, he announced, "Eric sleep in big bed"—and he did. Two weeks later, he helped his mom and dad dismantle the crib, and two weeks after that, assisted them in setting it up for his new sister, proudly announcing she could have it. He was a big boy. If your child isn't ready to give up his crib, forcing the issue is not going to help the relationship. Better to find another alternative for the baby.

Your child's nest is important to him. Unfortunately, he may communicate that message to you with his misbehavior rather than words. If he reacts to the change with defiance and tantrums, he's letting you know he's not quite comfortable yet. By taking the time to create a space that smells and feels right, and gradually introducing it to your child, you will help him discover that it is safe enough for him to sleep—without a fight.

Select the right bed for your child

As you create the new "nest" for your child, remember that you'll be coaching him, supporting him as he makes the move.

Toddler beds are an option When choosing the new "nest" for your child, think about how it will accommodate you. A toddler bed may be perfect for a child who feels most comfortable in a small space. It also allows you to use the same crib mattress that he may have been sleeping on already. The downside is that he'll soon grow out of it, and then you'll have to move him once more. And if your child likes to cuddle before he falls asleep, a toddler bed can be tough on you.

Loft beds serve a purpose A loft bed allows more floor space in a child's bedroom. If you prefer to stand next to your child rather than lie down with him, it provides a good excuse for you. And it's not easy to get out of a loft bed, so if you have a little wanderer, it may help to keep him there at night.

A few words of caution may also be in order, however. If your child is likely to attempt to crawl out on his own, or is highly active in his sleep, a loft bed may prove to be unsafe. Or, if he needs to get up in the middle of the night to use the bathroom or frequently wets, so you're constantly changing sheets, a loft bed may create more frustration than sound sleep. And if your child wants you to climb in with him, be sure you can.

A double bed provides space for you If you have the space, a double bed provides enough space for you to lie down with your child, if you choose. "Working full-time, I like our nighttime cuddles," Terri said. "We were ready to move our son out of our room, but I actually enjoy lying with him for a while. Unlike his older brother and sister, he's never needed a massage or music or anything. A few minutes of snuggling, and he's out. When we moved him to his own bed, we bought a queen-sized mattress, so that there would be plenty of room for both of us."

Once you've selected the type of bed your child will be sleeping in, take time to find the right bedding for her. In class, I asked the parents what they did to create a new "nest" for their child.

"I gave Avia the comforter off my bed. She likes the feel of it, and the smell of it even more. I was the one who got the new comforter instead of her."

"We used a weighted blanket and a body-size pillow."

"Jacob needed the bed placed away from the window."

"New sheets did it for Sara."

Work with your child, trying out different types of bedding, or transferring bedding from the old bed to the new. But don't stop there. Once the bed is right, make sure the room is, too.

Prepare the room If your child is eighteen months or older, have her help you prepare her new room. She can carry toys and favorite things to the new space. Select a motif that she enjoys; just be sure that the dancing bears look as friendly in the dark as during the day.

Spend time in the room Before you expect your child to sleep in the new room, spend time there. Read together in the space, play games. Visit during the day and in the evening. Talk about the sounds you hear. Show her the things she needs to feel comfortable while in the room, such as how to turn on the light, open and shut the door, or draw the blinds. Once again, I asked parents to share what they had done to make their child's room "feel right."

"Emma needed music and the door shut, but then she decided she liked it open better."

"The room had to be on the second level and next to us, for Dane to feel comfortable."

"Closet doors were most important for Avia. They had to be open so she could be sure no one was in there."

Laurie and Marc reported: "We offered Matt the guest room across the hall from us, but he chose to go to his own room farther down the hallway. We asked him how he wanted it to be, and involved him in the changes. That really made a difference."

By creating a comfortable "nest" and a setting for that "nest," you'll help your child to be calm enough to sleep.

Learning to sleep independently

Author Nino Ricci shared a bed with her mother for seven years. In her book *Lives of the Saints,* she wrote of the night her grandfather prepared a new room for her. "It was a room without a history, and my first night there, lying awkward and alone in my new bed, its air of abandonment seemed to hang over me like a pall."

When your child has been sleeping with you, it's important to recognize that the first skill he needs to develop is the ability to sleep without someone next to him. Before you "drop him" into a new bed, or into the room down the hall, it's essential that you gradually expose him to the experience of sleeping independently or with someone other than you.

Consider moving siblings in together The experience of sleeping alone in a bed, in one's own room, is very much limited to Western culture. It is a luxury rarely enjoyed or, for that matter, desired by many cultures of the world. In many societies, children are never expected to sleep alone. They simply move from their parents' bed to one shared with their siblings.

Your children may surprise you with how happily and well they share a room. Four-year-old Kate was going to lose her room to the new baby, but Mom and Dad didn't want to emphasize that. Instead, they told her, "a wonderful thing is going to happen. You get to move in with your brother!" She was delighted, and since both Kate and her brother were sound sleepers, there never was an issue of one waking the other.

Think about your children. Consider their ages, gender, and temperament. It may well be that they sleep better when they get to share their sleeping space. Laurel's three daughters enjoyed sleeping together so much that she transformed one of their rooms into the playroom, another into the study room, and the third became their sleeping room. If they needed it, they always had a space to go to, but in the middle of the night, they chose to be together.

In fact, it's not uncommon for individuals from large families or different cultures to say, "I never slept alone until I moved away from home."

When siblings are not an option If your child has slept with you, the sensation of sleeping alone can be alerting. So, if a sibling isn't available and your child is five-years-old or older, you may try the family pet. But if no sibling or pet is available, help your child become accustomed to independent sleeping by taking a gradual approach. Jason's family made it fun.

"Our three-year-old son had never slept alone, so, before we moved him into his own bed, we did a family 'camp out.' We all slept on the floor in our room in sleeping bags. Once he got used to that, we introduced his new bed."

If you wouldn't be comfortable sleeping in a bag on the floor, and your child is old enough that you are not concerned about smothering, put a rolled blanket or body pillow between you and your child. Your child will become accustomed to bumping up to the pillow or blanket rather than you. These items can then go with him to his new bed.

Try a mattress on the floor "Three-year-old Ben is a restless sleeper, so we got a mattress and put it on our floor. We'd lie down with him, and then we'd crawl into our own bed. If he awoke in the night, I held his hand, and he went back to sleep. Once he was comfortable on the mattress, we moved it across the room. Then we moved it down the hall."

Use naptime for practice It's often easier to practice new skills during the daytime. Naps can provide the first opportunity for practicing independent sleep.

"Zach slept with us at night, but I began to put him in the crib for his nap. Once he was napping well there, I started trying to do it at bedtime. Initially, he didn't like it. When he protested, I took him out again, but I kept trying every few nights. It took several weeks. One night, he just rolled over and went to sleep. I guess he finally got used to it."

Try a gradual "withdrawal" The Taylors had a king-sized "family" bed. Each of their three children actually had his or her own bed and bedroom, but all three kids preferred to sleep with Mom and Dad. Everyone loved it. But by the time the oldest was nine, it was getting a bit crowded, and Mom and Dad were ready for more privacy. They broached the subject with the kids, who responded with protests. The nine-year-old was especially upset by the idea. So, they offered the kids an alternative. Each child would be allowed to choose one night a week to sleep with Mom and Dad, and the remaining four nights would be "child-free."

Initially, Mom and Dad had planned to give them a week for the transition, but the youngest immediately claimed the first night as his. He liked instant gratification. The second-born picked Friday, knowing that she'd get more time and cuddles when everyone slept in on Saturday morning, and the nine-year-old, not to be outdone by his sister, chose Saturday night, to also extend his morning cuddles on Sundays. Their mom later laughed, "The youngest learned the days of the week by this arrangement. He was so eager to know when it was 'his turn' again that he'd count the days."

By gradually introducing to your child the concept of sleeping alone, you prepare your child for the experience. As a result, you avoid pushing him into a state of alert, leaving him wide awake and screaming in protest.

Provide support and time to practice

In his newspaper column, James Lileks wrote of his two-year-old daughter, who had just been introduced to her new "big bed." "She went down without any problem. Five minutes later I heard Louis Prima's "Yes, We Have No Bananas" coming from her CD player. I looked down the hall. She was standing there, clutching a stuffed bear. I put her back to bed, turned off the music, and asked if she would like any more of her 397 stuffed animals to join her. Five minutes later, I heard something outside my study door. The knob did not turn. Yet there were a pair of small feet visible under the door. Fine; wait her out. But after a while it got creepy; she was just standing there."

Becoming comfortable in a new sleep site and staying there is a process. You can save yourself a great deal of frustration if you expect to help your child learn to trust that he is "safe" in his new site. Even though you've talked about the move, created a "nest," and gradually introduced your child to the new environment, in order to keep him there, you can expect to spend some supportive practice time with him. So, select a time when you have four to eight weeks to work with him. Then start "practicing."

"I thought moving our son to a spacious new bed was a great idea," Hannah said, "but it was harder than I expected. He was quite distressed. He likes to lie with me. He says I protect him from the 'monsters' and 'bears.' So, we are making the shift gradually. I sit with him on his bed, and then kind of lean over him closely if he starts to get upset. It's taking time, but it's getting better."

"We transferred Sam into a crib by sitting next to him, holding his hand, until he fell asleep. He was probably eight months old. From there, we went to sitting in the room, and then sitting at the door, and finally, at fifteen months, he was in his own room, going to bed on his own and sleeping at least twelve hours through the night. Once he was comfortable and wouldn't disturb his brother, we moved them in together."

A helping hand as your child nods off, or knowing that you will respond in the middle of the night, allows your child to remain calm and discover that he really can sleep in his new site. By working with your child, and providing support, you help him to be successful. You stop the refusals and curtain calls before they ever start.

TIME: ENSURING YOUR CHILD'S BODY IS READY FOR SLEEP

While managing the tension is essential for sound sleep in a new site, the whole process is much easier if you also have Mother Nature on your side, telling your child's brain it's time for sleep—even in a different spot.

Maintain your routine

A routine keeps your child's body clock set, and clearly cues him that it's time to sleep. The more familiar your child is with the routine, the easier for him to fall asleep.

"Everyone told us we should shut the door and let her cry," Brenda lamented. "But you can't force Emily to sleep. It just doesn't work. A clear and consistent routine made the difference. I would bathe her, and then read to her and say our nighttime prayer. If she fought it, I didn't give in, but I also didn't force her. I simply said, 'Emily, I'm here. I will help you, but now it's time to sleep.' Following the same pattern was comforting to her, especially with the change. When she knew what to expect, she was more willing to comply and settle in."

By selecting a time when you can firmly stick to your routine, you will increase the chances that the change will flow much more easily. That's what Laurie and Marc discovered.

"Once we made the decision, we didn't go anywhere. We were on a mission this time. Staying put kept us on track. We didn't even have guests, knowing that we had to stay on our schedule. There was a goal to be achieved." If you maintain your routine, Mother Nature becomes your "backup."

Include extra soothing and calming activities

Even before you begin your bedtime routine, keep your evening calm and low-key. Plan to allow extra time for more soothing and calming activities as you move into your bedtime routine. Your child is in a new environment and, as a result, more likely to be on the edge of "alert." By keeping him in a state of calm tired, you make it much easier for him to fall asleep.

"I felt as though I had so much more influence over his sleep," Sue said. "I knew exactly what to do. While reading together, we sat in the rocker, and then I massaged him and did aromatherapy on his pulse points. He couldn't help but fall asleep."

TEMPERAMENT: CUSTOMIZING THE PLAN TO FIT YOUR CHILD

It's really the combination of temperament traits that can make shifting your child from one sleep site to another more challenging. The intense child alerts more easily when a situation is new, making it harder to sleep. The sensitive child notices what's different, and the slow-to-adapt child arouses with shifts or transitions. Combine them, and you'll work harder to help this child sleep.

When your child is intense

It's important to remember that when your child is intense, it's likely that he will protest the change. That doesn't mean he can't learn how to manage it. Understand that he's upset, and take the steps to help him work through it.

As I worked with Laurie and Marc's family, it quickly became apparent that Matt was a very intense child, so, despite the preparations and his involvement in them, initially, he still resisted. "I want a good night sleep. I need to sleep in your room," he pleaded. But Marc and Laurie responded clearly, "No, it's time for you to sleep here. We'll help you get through, by staying with you, but you need to sleep alone."

The first couple of nights he did wake up and come to get them. They returned with him to his room. After two weeks, he was only coming one night a week, and then Laurie decided one more little nudge was necessary.

"Six weeks into the process, I said, 'Matt, Mommy needs your help. I have to get my rest. You have to help me by staying in your room all night long.' I told him I would check on him before I went to sleep for the night, and then I left him while he was still awake. 'Remember, I need you to sleep all night,' I said once more. We could hardly believe it, but he did it and has ever since."

When you are working with your intense child, it is important to remember that "fighting" with him or leaving him to cry only fuels his intensity and increases his tension, making it harder for him to sleep. He needs you to work with him, supporting and helping him calm his body so that he can sleep, yet clearly communicating: it is time for sleep.

If you are intense, too, it's not easy, especially if you are tired. That's why it's so essential that when you begin the process, you clearly decide that you are ready. Then cut back on other commitments, and maintain a strict routine so that you have the patience and the stamina to help your child be successful.

When your child is sensitive

It's tempting, when changing your child's "nest," to entice your child into the move by buying a new bed, sheets, and/or blanket. Remember that the sensitive child is alerted by odors and knows the sensation of every lump in his mattress. If this child has been sleeping in your bed, you may consider giving it to him and enjoy being the recipient of the new items yourself.

When your child is slow to adapt

Your slow-to-adapt child does not like change and approaches new things cautiously. You may have already noticed that when he began to walk, he watched, waited, and then, one day, surprised you by standing up and moving. He had "practiced" with his eyes.

It's likely that you will be ready for a change long before this child is. Move slowly, introducing the idea and the new furniture. Provide support, taking time to play in the new space, and spend time there. Expect to wait. But know, too, that sometimes your slow-to-adapt child does benefit from a gentle nudge. By working together, you will get there, it just may be a little later than you anticipated.

When your child is high-energy

Safety is always a concern with a high-energy child. Be certain, as you change sleep sites, that you consider whether or not a high-energy child would be safe in this setting. Think carefully before putting him on a top bunk, even one with guardrails.

When your child's body rhythms are predictable and regular

Your child's predictable inner clock for sleep may prove to be an asset when you change his sleep site. If his routine is maintained, it's likely that this child will sleep anywhere!

When your child's body rhythms are irregular

The irregular child doesn't have the natural clock helping him know when to sleep, so, maintaining his routine and soothing and calming him will help him to manage the change more effectively. Without that support, he may end up short on sleep and less able to cope.

CELEBRATE SUCCESSES

Changing sleep sites for your child doesn't have to be a traumatic event. Instead, it can be an adventure that you experience together, a team working in tandem. By breaking the changes down into steps, you create a scaffold for success. There is joy and satisfaction in your accomplishments. Misbehavior rapidly diminishes as your child sleeps soundly in his new "nest," but, even more important, your relationship is strengthened. You and your child have experienced the pleasure of working together. It's a lesson that will follow you into the daytime hours, where you will discover that the difference between a child who is well rested and one who is not truly is a smile on his face—and on yours.

THE JOYS OF SOUND SLEEP AND GOOD BEHAVIOR

CONCLUSION

I wish I could tell you that, after reading this book, your family will always find it easy to get the sleep you need, that you will not experience any difficult choices, bedtime battles, or awakenings in the night, but I must be honest with you. I know that every day you will be faced with temptations—overnight party invitations, outings with friends, great sales at the mall, or favorite movies to watch. There will be classes, rehearsals, and practices to attend, opportunities you don't want your child to miss.

Becoming aware of the decisions that influence how much sleep your family gets, making sleep a priority, and figuring out the most effective techniques to help your child sleep is a process that continues over years and every stage of your child's life. Every step of the way you will be challenged in different ways to stop, take note, and decide. If today things don't go quite as you had hoped, tomorrow offers another opportunity and different choices. Allow yourself to gradually make it better. It's all right if there are fits and starts. The most important thing is to be thinking about sleep, planning for it, and striving to get it. Some days will be better than others. And when you realize that you've "lost the way," come back and start again. It will be easier the next time around.

It's said that it only takes ten percent of the population to begin a social change. Perhaps this is a beginning. One by one, we will commence to brag about how much sleep our children are experiencing, letting it be a source of pride, a badge of honor; the ticket to truly maximizing their potential.

Until that day arrives, however, I suspect you may sometimes feel that you are the only parent going home for naps or saying no to a "second sport." There will be days when you wonder if you are wasting your time helping your child to unwind after a hectic day when others don't and tell you that you shouldn't. And nights when you get discouraged because, once again, your sensitive sleeper needs more help settling for the night. When those feelings strike, pick up this book and read it again. Let it be your guide, a helping hand in the middle of the night. Listen to the voices of other parents who have shared similar troubles and tribulations. Review the information, and unearth the "hints" that you have previously skimmed over, which may now prove to be very useful. Make them your own.

And, as you help your child to sleep, learn about your own needs. Recognize the most effective strategies for reducing the tension in your body. Identify the key events that set your body clock. Become aware of the little things, which really do allow you to easily fall asleep and slip back into it again—even when you are awakened in the night.

Don't forget to stop and celebrate your successes. Too often, we don't even realize the toll exhaustion is taking on us. Emotions become dull, life gray, existence a challenge. So, rejoice in the "good" nights. Give yourself a pat on the back when you make the tough decision to honor your child's bedtime. Acknowledge the strength it has taken to do what you know in your heart is in the best interest of your child's well-being—even if it is an unpopular choice. Applaud the resulting good behavior your child demonstrates during the day, and your own astute skill as a parent.

Something mysteriously wonderful happens as you and your child truly get the sleep you need. Life becomes a hot-fudge sundae, sweet, delicious, and irresistible. Sound sleep catapults you into the day ready to consume every last ounce of it. You are powerful, enticing, smart, and capable when drenched in sleep. Happiness glows in your eyes and skin. It resounds in your laughter, deep and infectious.

Your children respond accordingly. Everything is easier; they listen and cooperate, and actually seem to enjoy one another's company. Surprises are exciting instead of daunting. And, sometimes, they even offer to help without being asked.

A good night's sleep has a power of its own. It allows you to truly discover the delight of living in a world that never stops—yet is enjoyed most—when you do. Choose sleep!

INDEX

accidents, 11, 22, 23, 33
ADHD, lack of sleep mistaken for, 13
adolescents
average sleep needs, 3, 123
bedtime routine for, 178
change to later wake time, 109
naps for, 241–42
sleep-deprivation in, 3
aggression and hitting, 10, 22, 23, 26, 32, 33, 247
allergies and asthma, 225
Anders, Tom, 80, 253
anger, 31, 33
anticipation and sleeplessness, 79–80, 83
anxiousness, 10, 21, 25, 32, 98–99, 194, 201–4, 212
attention, focus, and performance, 11–12
checklist for children, 24–25
checklist for parents, 34
awakening, morning, 124–27
alarm clock or sudden, 32
body rhythms irregular and, 277
clocks for your child, 275
difficulty with, 22
easing tension in the morning, 89–92
establishing a wake time, 124–26, 158
goal wake time, 126–27
high-energy child, 277–78
intense child, 276–77
irregular wake times, body clock disruption and, 107–9
moving wake times earlier, 278–81
moving wake times later, 268–78
sensitive child, 277
slow adapter, 159, 277
snacks for, 275
toys for, 274–75, 277, 278
waking up your child, 90, 124–25

bath time, 17, 177, 194–95
"tricking" the brain with, 279–80
bed, changing of (and changing rooms), 296–315
abrupt change, how to handle, 301
bedding, types of, 304–5
body rhythms and, 312
celebrating success, 312
creating a new "nest," 302–3, 304
crib, giving up, 303
customizing to fit your child's temperament, 310–12
family pet for comfort, 306

bed (*cont.*)
gradual move to sleeping alone, 306–8
growth spurts and, 300
high-energy child, 312
intense child, 310–11
learning to sleep independently, 305–6
maintaining your routine, 309
mattress on the floor, 307
naptime for practice, 307
new baby and, 300
as parental decision, 298–99
parents spending time in the room, 305, 306–7
preparing the room, 305
providing support and time to practice, 308–9
selecting a bed, 304–5
sensitive child, 311
sharing the idea with your child, 301–2, 305
siblings sharing a room or bed, 306
slow-adapting child, 311
soothing and calming activities, extra, 309–10
steps to take, 297–98
tantrums, defiance, and, 303
"window" for change, 299–300
bedtime
body rhythms, honoring and, 160
customizing to fit your child, 188–204
customizing to fit your infant, 263–65
debates, avoiding, 184
establishing clear limits, 202–3
establishing regular, 133–34, 158
fearful child, managing, 201–4
four step routine (TCCS), 176–85
infants, 262–65
irregular, sleep problems and, 109–10, 161–62
managing callbacks, 197, 202
moving bedtimes earlier, 278–81
moving bedtimes later, 268–78
predictability, prioritizing, 159, 175–76, 196, 198, 202
props for, 98, 179–80
routines, 17, 55, 134, 167–87, 262–65
sample routines, 185–87
seeing potential trouble spots, 171–75
sleep time vs., 128
staying up past, 14, 23, 103, 105
touch eases tension, 98–100
visual planner for, 184–85
wildness at, 22
wind-downs, 81, 128, 133–34, 152, 158, 169–70
See also environment for sleep; strategies for sleep
Belsky, Jay, 259
Beltramini, Antonio, 208
body clock, 14, 103–43
adjusting for morning lark, 127
adjusting for night owls, 127–28
average sleep needs for each age group, 3, 123, 236
bowel movements and, 247, 270
chart of daily activities that affect sleep, 106–7
daily routines, 107–13, 159, 161–62, 219–20, 270–73
decisions daily and confusing or throwing out of balance, 103–4, 119
determining your child's natural sleep time, 128–29
determining your child's natural wake time, 124–26

goal for wake time (resetting clock), 126–27
how it works, 103
infants, 253
irregular bedtimes, 109–10
irregular body rhythms and, 160–62
irregular mealtimes, 110–11
irregular naptimes, 111–13
irregular schedule and, 107
irregular wake times, 107–9
jet lag and, 104–5
lack of exercise or exercise at the wrong time, 107, 115–17
light and, 107, 113–15, 136–37, 220
moving wake and bedtimes earlier, 278–81
moving wake and bedtimes later, 268–78
night waking and, 219–21
resetting, time needed to move, 236–37, 268
routine vs. rut, 121–22
seasonal changes, 129
sleeping in and, 108, 109, 127
staying up past bedtime and, 14, 23, 103, 105
stimulants and, 107
travel and, 288–89
upset of, as trigger of tense energy, 65
See also window for sleep
body control and behavior, 3, 9, 10–13
alert mode (fight or flight), 33, 63, 64
checklist for children, 22–23
checklist for parents, 32–33
intense temperament and, 150
movements indicating fatigue, 22
response to various sleep strategies, 52
body rhythms, 160
naps and, 246–47
sleeplessness and irregular, 160–62, 197–98
temperament and, 160–62, 197–98, 246–47, 277, 294–95, 312
travel and, 294–95
bossiness, 25, 32
breakfast, 90–92, 270–71, 279
Brooks, Robert, 115

caffeine, 14, 102, 103, 117–18, 136, 236, 289
calendar, marking of, 209–13
adult-only vacations or business trips, 212
Daylight Savings Time, 278–79
growth spurts, when to expect, 209–11
holidays, 211
minimizing commitments, 212–13
upsetting events and anniversaries of, 211–12
calm energy (green zone), 25, 62, 172–73
connection and, 94–95, 193
creating, 89–101
creating in intense temperament children, 152, 191–93
calm tired, 62–63, 172
Cambridge Friends School, 231–32
Cameron, Jim, 152, 246
carbohydrate cravings, 10, 33, 185
chart of daily activities that affect sleep, 106–7
Chervin, Ronald, 24
Chess, Stella, 146–47
child's room. *See* environment for sleep
Churchill, Winston, 232
clumsiness, 11, 22, 23, 33
competition and pressure to perform, 71, 81, 83

computer. *See* television, computer, and radio
connecting and calming activities, 176, 178–81, 199
  bed or room change and, 309–10
  calm energy and, 94–95
  calming ritual, 180–81
  cue for sleep, 181–82, 244
  individual differences, 178–79
  lovies, 179–80, 199, 214, 224, 236
  naptime routine, 243
  night wakings and, 214–15
  rituals of, breakfast, 90–92
  special moments through the day, 92–93
  switch to sleep and, 182–84, 215
  for travel, 285–86
Cooke, Betty, 49
cortisol. *See* stress hormones
Covey, Steven, 95, 122
crying
  as "cues," recognizing, 47–48, 54
  falling asleep and, 47
  ineffective sleep strategies and, 48
  infants and sleep, 59–61, 262–63
  intense children, importance of not leaving them to cry, 150–51, 222–23
  let your child know you will respond, 215
  weepiness, general, 21, 31
cue for sleep activity, 176, 181–82, 244
  infants, 262–63
Czeisler, Chuck, 113

Dahl, Ron, 45
decisions
  daily, affect on sleeping, 14, 16–17, 102–6, 119
  -making, fatigue and difficulty with, 11–12, 24, 34
Dement, William, 29
depression, 29
Dutch children
  average amount of sleep, 239
  focus on children's rest in the culture, 7
  fussy infants vs. U.S., 258

Edison, Thomas, 231
emotions
  checklist for children, 20–22
  checklist for parents, 31–32
  comforting distress, 215–17
  difficulty managing and sleep deprivation, 10, 18
  distress and excitement, 70–84
  distress and excitement checklist, 82–83
  "I" messages and, 88
  overload, 21
  parental responses to child's behavior and, 46–47
  phrases to identify feelings for your child, 87–88, 292
  sleep sensitive to, 70
  as trigger of tense energy, 65
  "volcano" analogy, talking with your child about feelings, 85–97
  worry journal for, 194, 212
environment for sleep, 14
  bed and/or room, changing, 296–312
  bedding, 304–5
  bed size and type, 224, 304–5
  child-proofing, 274
  clock in child's room, 275
  colors, 97
  co-sleeping and related options, 216, 255–56, 307
  creating a "nest" for sleep, 97–98, 155–56, 245, 302–3

dark-out blinds, 140
infants, 253–58
light, 107, 113–15
limit visual clutter, 195–96
night lights, 140
parent sleeping in child's room, 275
safety and security, 98, 182–83
slow adapters and, 223–24
temperature, 97
white noise for, 195
See also bed, changing

exercise
body temperature and switch for sleep, 116, 141, 271, 280
falling asleep easier and, 280–81
late afternoon, to move bedtime later, 271–72
scheduling, 140–41, 163
sleeplessness and lack of, or at wrong time, 107, 115–17
travel and holidays, planning, 290–91

falling asleep
cues, establishing, 176, 181–82, 244, 262–63
difficulty with, 7–8, 11, 15, 24, 32
dimming lights, 140, 158, 243, 245, 262, 262
exercise and, 140–41
gradual descent into sleep, 168–69
green zone of calm tired and, 62–63
instilling a sense of security and, 45
red zone of tense tired and, 62, 65
wind-downs, planning for, 81, 128, 152, 158, 169–70
"window" for sleep, catching, 15, 110, 130–33, 235–36, 238–39, 240–41
See also bedtime; naps; strategies

family sleep schedule
adjusting window for sleep and, 132–33
average hours needed by age group. 123
cue for sleep and family history, 181–82
goal schedule, 143
goal sleep/activity chart, 144
moving wake and bedtimes earlier, 278–81
moving wake and bedtimes later, 268–78
naps for, 242
prioritizing sleep, 142–43
scheduling members for sleep, synchronizing, 15, 123–24, 143, 267–68

fears, managing bedtime, 201–4
Flinn, Mark, 72–73
forgetfulness, 11–12, 24, 34, 96
Franklin, Benjamin, 209
frustration levels, 20, 25, 31

Gelula, Richard L., 289
goals for promoting sleep in children
goal sleep/activity chart, 144
moving wake and bedtimes earlier, 278–81
moving wake and bedtimes later, 268–78
for newborns and young infants, 252
night waking, reducing, 227–28
opportunities for intimacy for adults, 15
schedule, 143
sensitive, responsive care, 15
sleep for everyone, 15, 185
structure for child to fit into family's way of sleeping, 15, 123–24, 143

Gottman, John, 48, 94
growth spurts and developmental milestones, 71, 80–81, 83, 209–11
  changing beds and, 300–301
  increased appetite and night snacks, 213–14
  naps and, 239–40
  night wakings and, 212–14
Gunnar, Megan, 45, 74, 222

Harkness, Sara, 7
Hayes, Marie, 251
headaches, 10, 21, 32
Heart of Parenting, The (Gottman), 48
holidays
  anticipation and sleep problems, 79, 211
  travel and, 283–95
  See also travel
How to Negotiate with Kids (Brown), 89–90
hyperactivity (wildness, frenzied behavior), 3, 9, 10–11, 13, 22, 23, 26, 32, 61, 64, 190
hypersensitivity (sensitive sleepers), 59–61, 65, 147, 153–55, 188–89
  find the right touch, 194–95
  give your child words, 193–94
  limit visual clutter, 195–96
  talk about emotions, 194
  white noise for, 195
  See also temperament, sensitive child

illness and immune function, 11, 22, 23, 33
  night waking and, 224–25
  sleep to counter, 96
ILLS, 107
infants, 251–66
  amount of sleep needed, 3, 6, 123, 252–53
  body clock, 253
  burp cloth or blanket, 257, 258
  cuddling, need for, 234, 263
  cueing for time to sleep, 262–63
  fussy, 258
  goals for first months, 252
  growth spurts, 210, 239–40
  high energy baby, safety and, 265
  how to get baby to sleep, 258
  intense temperaments, strategies for, 151, 264
  massage, 261–62
  nap for, 232, 238
  nap window, finding, 238–39
  newborn, 251–52
  night cap for, 261
  nightgowns for, 261
  parental schedule and, 263
  parental stress and reestablishing intimacy, 258–60
  preparing infant for sleep, 59–61
  problem nappers or leader babies, 233–35
  room change and, 74
  sensitive baby, 59–61, 65, 153, 154–55, 183, 263–64, 265
  shhh, shhh, shhh sound and, 262
  signs of stress in, 3–4
  sleep-deprivation in, 3
  sleeping arrangements, "window" for change, 300
  slowing down for, 234
  stages of sleep in, 206
  Sudden Infant Death Syndrome, 260
  swaddling, 260–61
  thumb-sucking, 261
  traveling with, 257

wake times, 109
weaning and calming activities, 180
what's normal, 252–53
where a baby should sleep, 253–58
irritability (crankiness), 9, 12, 21, 25, 29, 31, 236, 240, 289

"jet lag," 14, 104–5

Kennedy, Pam, 261
Kids, Parents, and Power Struggles (Kurcinka), 5
Kübler-Ross, Elisabeth, 102
"kur" technique, 273

Leksand, B. Eckerberg, 224
life changes
as distress trigger, 71, 76, 82
slow adapters and, 157–60
See also bed and/or room, changing; travel and holidays
light and sleep, 107
body clock and, 113–15, 288
dimming in the afternoon, for naps, 243, 245
dimming in the evening, 140, 158, 262
exposure to, importance of morning, 136–37, 288
morning, 113, 115
moving to later bedtime and keeping on, 272
seasonal changes, 129
sensitive children and, 155
wrong time, from computer or TV screens, 113–15
listening
children's cues, hearing, 47–48, 55
difficulty with, 5, 11–12, 24
diffusing tension and, 93–94
Lives of the Saints (Ricci), 305
lovies, 179–80, 199, 214, 224, 236

Mahowald, Mark, 230
Mass, James, 107
massage therapy, 98–100, 133–34, 214
"batch of brownies" massage, 99
infants, 261–62
McKenna, James, 256
meals
breakfast, 90–92
final, before bedtime, 136
irregular schedule and sleep problems, 110–11
schedule for, 136
misbehavior, xv, 3–4, 7–8, 16–17, 18–19, 27
associated with missing sleep, checklists, 20–26
body clock upset and, 105
parental stress and, 73
power struggles, 4, 6, 20, 29–30, 37, 102
signs of sleep deprivation and, 9–13
tantrums, xv, 4, 7–8, 16–17, 18, 19, 22, 161, 211, 226, 231, 303

naps, 111–13, 229–48
adolescent naps, 241–42
average needs by age group, 134–35
averaging sleep time and, 129, 236
bedtime and, 236
behavior on no-nap days, 231
benefits for all ages, 230–31
body clock and, 124
body clock, resetting and, 271
body rhythms regular and predictable and, 246–47
bowel movements and, 247
celebrating success, 247–48

naps (*cont.*)
change in and stress, 74, 279
childcare centers, timing and, 243
common culprits behind nap resistance, 236–38
creative approach to, 232–33
Daylight Savings Time and, 279
essential nature of, 230–31
family naps, 242
finding your child's nap "window," 235–36, 237, 238–39, 240–41
growth spurts and, 239–40
implementing, 235
infants, 232, 238–39
intense temperament children, 150, 245
late afternoon, problems with, 103, 112
night terrors and, 226
night waking and, 220
parental expectations, reviewing, 237
problem nappers or leader babies, 233–35
regular times for, 17, 134–36, 159, 229, 242–43, 246
resistance to, 10
reintroducing, 245
sensitive temperament children, 245–46
skipping, 14, 103, 111, 112, 220, 248
slow-adapting children, 246
as tension buster, 241
toddlers, 240–41
travel and, 288–89
what they look like, 231–32
when your child wants you to nap, 233–35
working together on, 237–38
naps, routine for, 242–48
connect and calm activity, 243
cue for sleep, establishing, 244
for high-energy child, 247
for intense child, 245
for irregular body rhythm child, 246
for sensitive child, 245–46
for slow adapter, 246
transition activity, 242–43
troubleshooting: working with the nap-resister, 243–44
switch to sleep, 244–45
National Sleep Foundation, 8, 164, 291
Newman, Mary, 231–32
night clothing, 155, 261, 286
nightmares, 226–27
night terrors, 225–26
night waking, 17, 205–6
body clock upset (jet lag), 219–21
environmental changes, 221
growth spurts and, 209–11, 212–14
holidays and, 211
infants, 253
medical issues, 205, 224–25
nightmares, 226–27
night terrors, 225–26
as normal and expected, 208–9
temperament and more frequent, 221–24
tension and, 207–19
why we wake, 206
night waking, strategies for
avoiding doubling up on tension-producing events, 213
bathroom trips, 218–19
check your bedtime routine, 221
comforting distress, 215–17
intense children, 222–23
let your child know you will respond, 215, 222–23
light exposures, 220

mark your calendar, 209–13
minimizing commitments, 212–13
more soothing, calming activities, 214–15
parental bed and, 216
sensitive children, 223
siblings together, 217–18
sleeping bag strategy, 216–17
slow adapting children, 223–24
teaching coping strategies, 219
work together with your child, 218–19
noise, exhausted child's desire for, 11, 24, 77
Novosad, Claire, 121, 150

occupational therapist, 156
Osgood, Charles, 167
OTC (over-the-counter) drugs, stimulants in, 291
Ounce of Prevention, 152, 246
overscheduling of children, 7–8, 78–79, 83
intense children, strategies for, 151–52
observing your child and, 95–96
slowing the pace, 95–96
vacations or travel and, 289
Owens, Judith, 26

Paidea Child Development Center, 232, 242, 245
parents
advice (ineffective) given to for getting child to sleep, list, 41–43, 46
attitude and changing perspective about child's sleep problems, 36–37, 58
average sleep needs, 123
becoming a sleep manager, 36–37
bed, co-sleeping and, 216, 255–56, 307
business travel and child's distress, 74, 82
celebrating success, 54–55
changing child's bed or room, as parental decision, 298–99
changing habits, 15–16, 105–6
checking your level of tension and fatigue, 171–72
decisions that prevent sleep in children by, 14, 16–17, 102–6, 119
evaluating your sleep-deprivation, 31–36
getting your own sleep needs met, 185
goals for sleep, 15, 52–53
guidelines, general, for getting children to sleep, 44, 48–52
infants and, ways to ease parental stress and reestablish intimacy, 258–60
life partner, keeping connected with, 94–95
life partner, sleep problems of, 208–9
naps for, 230–31
night awakenings, 96, 205
responding to advice givers, phrases to use, 53–54
sleep-deprivation in, 4–6, 28–31, 33, 46, 283, 284
slowing down, need for, and "leader babies," 234–35
spending time in child's new room, 305, 306–7
stress level and child's distress, 71, 72–73, 82, 90, 194
traveling with children, 283–95
uniqueness of individual children, recognizing, 43–45

parents (*cont.*)
waking up kids, 90, 124–25
window for sleep, identifying yours, 133
See also strategies for getting children to sleep
Periodic Limb Movement (PLMS), 164
Perry, Susan, 129
picture planner
of bedtime routine, 184–85
of moving to new awakening or bedtime, 285–76, 280
of travel and holiday plans, 292
Porges, Steven, 181
power struggles, 4, 6, 20, 29–30, 102
preschooler
average sleep needs, 123, 236
growth spurts, 210–11
naps, 231–32, 237–38
nightmares, 227

questions to end your day, 100–101

Raising Your Spirited Child (Kurcinka), 5, 20, 164
Raising Your Spirited Child Workbook (Kurcinka), 5, 164
relaxation nights, 99–100
Restless Leg Syndrome (RLS), 163–64
Ricci, Nino, 305
routines, daily
bed or room changing, maintaining routine and, 309
bedtime, 17, 55, 134, 167–87
changes in, 20, 31
changes in, slow-adapting children and, 157–62
creating and keeping, 161–62
Daylight Savings Time and, 272, 273, 278
importance of regular and predictable, 107–13, 159, 161–62, 237, 246
for infants, 258–65
meals, 136
moving wake and bedtimes earlier, 278–81
moving wake and bedtimes later, 268–78
naps, 17, 134–36, 159, 220, 237, 242–48
night waking, reviewing schedule and, 219–20
travel and maintaining, 288
rule-breaking, 22, 25

Sadeh, Avi, 12, 221
safety and security, children's need for, 45, 63, 94, 182–83, 226, 244
bed or crib and, 299
sensitive children and, 156
school
ADHD, lack of sleep mistaken for, 13
anticipation and sleeplessness, 79–80, 83
competition and pressure to perform, 71, 81, 83
homework, 21–22, 71–72, 81
inability to focus or perform well, 11–12, 24–25
studying late or sleeping, results, 25
upsetting events in, 75–76, 82
school-age children
average sleep needs, 3, 123
naps for, 232
night wakings, 208
sleep-deprivation in, 3
Schwarz, Norbert, 185
"secrets" of sound sleep, xvi, 14–15
sensory integration disorder, 156

separations as distress trigger, 71, 73–75, 82, 212
separation anxiety, 25
Sheldon, Stephen, 10–11
sleep
average needs by age group, 3, 123, 130
benefits, 96
crash, 62, 64–65, 109
debt, 12, 23, 215
family sleep schedule, 123–24
planning for adequate hours, 129–30
prioritizing needs, 122–24, 142–43
restorative, 17, 62, 63, 288
schedule, 107
seasonal fluctuations, 129
See also body clock; falling asleep
sleep apnea, 204
sleep-deprivation
accumulated, 12
amount of sleep needed for all age groups, 3, 123, 130
attention, focus, and performance, 24–25, 34
body control and behavior, 3, 9, 10–13, 22–23, 32–33
causes and lifestyle choices, 7–8
discovering your child is missing sleep, 8–9
emotions and, 10, 18, 21–22, 31–32
evaluation of your child, checklist, 20–27
evaluation of yourself, checklist, 31–36
extent of problem in America, 13–14
illness and, 11, 22
intense temperament and, 151–52
misbehavior and, 3–4, 7–26
mood and, 28–31
pain (headaches, stomachaches), 10, 21, 32
recognizing in yourself, 4–6, 31–36
signs of, 9–13, 20–26, 31–36
social situations and, 12–13, 25–26, 35–36
as trigger of emotional distress, 71, 76, 82
sleep journal, 120
establishing a morning wake time, 125–26
sleeplessness, 57–66
fearfulness and, 201–2
high-energy children and, 162–63, 198–99, 224
high intensity temperament children and, 150, 189–93, 222–23
high-sensitivity children and, 153–55, 193–96, 223
in infants, 59–61
in older children, 61
insecurity and hypervigilance, 45, 61
irregular body rhythms and, 161, 197–98
irregular schedule and, 107, 158
lack of exercise or exercise at the wrong time and, 107, 115–17
light exposure, 107, 113–15, 220
medical problems, 204
sleep deprivation as cause, 108–9, 129
slow adapters and, 158–60, 196–97, 223–24
stimulants and, 107
stimulation and, 7–8, 11, 15, 24
tense energy (tension) and, xv, 62, 64–65, 69–84
unexplained or mysterious wakefulness, 200
vulnerability, feelings of, 200–201

snoring, 225
social skills, 12–13
  checklist for children, 25–26
  checklist for parents, 35–36
Spiegel, Karine, 23
stimulants. *See* caffeine
stimulation and overstimulation, 71, 77–78, 107
  checklist for evaluating, 83
  desire for loud noise, 11, 24, 77
  intense temperament, slowing pace for, 151–52
  light exposure, 113–15, 137, 140, 195, 220
  overscheduling of children, 7, 71, 78–79, 83, 95–96
  protecting child from, 15
  roughhousing or vigorous activity before bedtime, 103, 116–17, 194
  sensitive temperament, reducing stimulation for, 155, 194–96, 223
  slowing daily pace as anecdote, 95–96
  symptoms, 77, 79
  television and computers, 75, 82, 91, 103, 137–39, 195
  wind-downs, planning for, 81, 113–15, 128, 152, 158, 169–70
stomachaches, 21
strategies for getting children to sleep
  bedtime routine, four-step, 176–85
  body's response to various approaches, 52
  brain's "switch" to sleep, 58–66, 110
  children's cues, understanding, 47–48, 54, 55, 170
  creating a "nest" for sleep, 97–98, 155–56, 254–55, 257
  effective, 51–52
  evaluating level of tension and fatigue, 172–73
  exercise, 163
  flexibility in, 15–16
  general guidelines, 44
  ineffective, 41–42, 45–47, 50–51
  infants, 262–65
  light, controlling, 113–15, 140, 158, 243, 245, 262, 262
  managing high-energy children, 163–64, 198–99
  managing intense temperaments, 150–52, 189–93
  managing irregular body rhythms, 160–62, 197–98
  managing sensitive temperaments, 154–55, 193–96
  managing slow adapters, 158–60, 196–97
  predicting trouble spots, 189
  reducing choices, 201
  safe, secure feeling and, 45, 55, 97
  structuring for success, 48–49
  successful, 54–55
  touch eases tension, 98–100, 182
  working together, 173–76
  See also night wakings, strategies for
stress hormones (cortisol) and sleeplessness, 47, 62, 63, 73, 74, 109, 271
Sudden Infant Death Syndrome, 260
"switch" for sleep, 14, 58–62
  babies, preparation for sleep in, 59–61
  body temperature and, 116, 141, 271, 280
  calm energy and calm tired (green zone) 62–63, 94, 152
  calming ritual, cue for sleep, and, 180–84, 244
  cues, 170, 244, 262–63
  irregular bedtimes and, 110

light and, 113–15
naps, 244–45
night wakings and, 214–15
pre-bedtime activities and, 117, 169–70
reverse, for awakening, 89–92
switch activities, 176–77

talking, excessive, 24, 37
tantrums, xv, 4, 7–8, 16–17, 18, 19, 22, 161, 211, 226, 231, 303
television, computer, and radio
evening watching, and lack of sleep, 103
light from and disruption of body clock, 113–15, 137, 220, 265
limiting screen time (turning off), 91, 137–39, 195, 265
loud, 11, 24
news on, upsetting, 75, 82, 211–12
temperament, 145–64
adaptable child, slower, 156–60, 196–97, 223–24, 246, 277, 293–94, 311
body rhythms and, 160–62, 197–98, 246–47, 277, 294–95, 312
changing beds and/or rooms and, 310–12
customizing bedtime routine for differences, 189–200
customizing naptime routine for differences, 245–48
customizing travel for differences, 292–95
environment or parental nurturing and influencing, 147
high-energy child, 162–64, 198–200, 224, 247, 294, 312
intense child 149–52, 189–93, 222–23, 245, 276–77, 292–93, 310–11
night waking and, 221–24
reviewing your child's temperament picture (scoring), 164
sensitive child, 65, 117, 152–56, 188–89, 193–96, 223, 245–46, 277, 281, 293, 311
spirited children, 146
what it is, 146–47
working with your child's temperament, 148
tense arousal, 62
tense energy (tension, red zone), xv, 62, 64–65, 69–84
catch and connect, 190
checking your and your child's level of tension and fatigue, 171–73
checklist for evaluating, 82–83
child's level of, 72
cues for, 191–92
cumulative effect, 83–84
easing tension in the evening, 97–101
easing tension in the morning, 89–92
easing tension throughout the day, 92–96
holidays and travel, 211, 284–95
hunger (irregular mealtimes) and, 111
infants, managing, 258–65
managing as team effort, 88–89
moving wake and bedtimes earlier and, 278–79
moving wake and bedtimes later and, 268–70
night waking and, 207–19
planning tension busters, 191, 241
phrases to identify feelings for your child, 87–88
siestas for, 241

tense energy (*cont.*)
soothing your child, 87, 215–17
sleep environment and, 97–98, 274
slowing down and, 234, 279
symptoms of, 64, 172, 173
touch eases tension, 98–100, 182, 214
triggers of, 65–66, 70–84, 108–9
"volcano" analogy, finding a way to talk with your child about tension, 85–97
wind-downs for, 81, 128, 133–34, 152, 158, 169–70, 192–93
working together on, 190–91
tense tired, 62, 215
Thayer, Robert, 62, 63
Thomas, Alexander, 146
Thomas, Ruth, 49
thumb-sucking, 261
toddlers
amount of sleep needed, 3, 6, 123
childcare and nap times, 243
childcare and stress levels, 74, 82
growth spurts and sleeplessness, 80, 83, 210
nap for, 232, 238
room change and, 74 (*see also* bed or room, changing)
sleep-deprivation in, 3
stages of sleep in, 206–7
travel and, 296–87
window for sleep, naps, 240–41
Tolan, Stephanie, 163
touch, 98–100
for infants, 261–62
for intense child, 151
for nightmares, 227
for night wakings, 214
parental, reestablishing intimacy, 258–60
switch to sleep and, 182
transition activity
bedtime routine, 176, 177–78
nap routine, 242–43
slow adapters, 223–24, 277
travel and holidays, 283–95
adjusting your child's body clock, 288
anticipation and sleep problems, 79, 211
body rhythms, irregular and, 294–95
body rhythms, regular, and, 294
business or adult vacation and child's distress, 74, 82, 212
celebrating success, 295
childproofing environment, 287
creating travel rituals, 285–86
Disney and, 289
exercise during, 290–91
high-energy child and, 294
hydration during, 291
infants, 257
intense child and, 292–93
lights out in one room, 286
naps and siesta time, 288–89
over-the-counter drugs, disruption of sleep and, 291
packing for your child, 285–86
parental stress of monitoring children in strange places, 286–87
planning, fitting your family's body clocks, 284–85
reentry time back home, 187
relatives, sleep schedules and, 290
sensitive child and, 293
slow-adapting child and, 293–94
working together, 2982

upsetting events, 71, 75–76, 82, 200–201, 211–12
anniversaries of, 211
night terrors and, 226

Van Cauter, Eve, 23
Visuals for Learning picture planner, 280
"volcano" analogy, finding a way to talk with your child about tension, 85–97

water and hydration, 291
  bottled, for sensitive child, 293
whining, 9, 10, 16, 19, 28
white noise, 195, 245
wildness. *See* hyperactivity
wind-downs, 81, 128, 133–34, 152, 158, 169–70, 192–93
"window" for sleep, 15, 110, 130–33
  adjusting, 132–33
  cues for identifying, 131–33
  intense children and, 245
  missing and time between windows, 131, 183
  moved, 197
  nap window, 235–36, 238–39, 240–41
worry journal, 194, 212

BOOKS BY MARY SHEEDY KURCINKA

SLEEPLESS IN AMERICA

Is Your Child Misbehaving or Missing Sleep?

ISBN 0-06-073602-X (paperback)

A practical guide to understanding the link between behavioral problems and sleep deprivation, along with a five-step process to help children get a good night's sleep.

RAISING YOUR SPIRITED CHILD

A Guide for Parents Whose Child is More Intense, Sensitive, Perceptive, Persistent, Energetic

ISBN 0-06-073966-5 (paperback)

Offers parents emotional support and proven strategies for handling their spirited children. Filled with personal insight and authoritative advice, *Raising Your Spirited Child* can help make parenting the joy it should be, rather than the trial it can be.

RAISING YOUR SPIRITED CHILD WORKBOOK

A Companion Workbook to Help Parents Create Families Where Spirit Thrives

ISBN 0-06-095240-7 (paperback)

This companion workbook offers parents and educators insights, emotional support, and proven strategies for dealing with spirited children.

KIDS, PARENTS AND POWER STRUGGLES

Winning for a Lifetime

ISBN 0-06-093043-8 (paperback)

Kurcinka offers unique approaches for solving the daily, and often draining, power struggles between you and your child. You'll be able to identify the trigger situations that set off these struggles and get to the root of the emotions and needs of both you and your child.